LEADER

VR & THE METAVERSE CHURCH

HOW GOD IS MOVING IN THIS VIRTUAL, YET QUITE REAL, REALITY

JEFF REED

LEADERSHIP NETWORK

VR & the Metaverse Church: How God Is Moving in This Virtual, Yet Quite Real, Reality

Leadership Network is a community of multipliers who gather to collaborate, innovate, and pursue what God has next for his Church. Our mission is to champion healthy church growth that is capable of reproducing. For more information, visit leadnet.org.

This book is manufactured in the United States.

ISBN: 978-1-959844-00-6 (print)

ISBN: 978-1-959844-01-3 (ebook)

Edited by Karen Cain
Cover by ArtSpeak Creative
Interior design by Karis Pratt

ACKNOWLEDGEMENTS

To my parents John and Linda: thanks for being supportive over the years, even when you didn't understand what you were supporting.

To my family—Amy, Caitlyn, and CJ: thanks for being my physical anchor and for giving me permission to do things never done before.

To James Srodulski, David Parks, Toney Upton, and Ray Raley: thanks for thinking the crazy ideas I had in my youth were not that crazy.

To Jim Tomberlin and my RESI friends: thanks for giving me a platform when no one else would.

To Chestly Lunday and Chris Caputo: thank you for the bandwidth to carve out new paths.

To the island of misfit toys we call Digital Church Network: it's an honor to support your dreams and calling.

To the leadership of NewThing Network, Leadership Network, and Exponential: thank you for seeing an opportunity to be the Church, to multiply disciple makers and movement leaders differently. In Dave Ferguson we trust.

To those pioneers and disrupters in virtual spaces: my prayer is that through this book you will be seen, heard, and understood. I am humbled by your tenacity for the Kingdom.

To those inspired by the following ramblings: my prayer is that God would do more through you and your missional imaginative ideas than you ever dreamed possible, and that you too can look back at your life one day in realization that only God could have gotten you to this point.

To the Goers, keep following God . . . even into the virtual realms.

CONTENTS

	Acknowledgements	iii
Intro	Church, Meet the Metaverse	1
1	What Exactly Is the Metaverse?	3
2	Virtual Reality 101	13
3	The Church in Virtual Reality	27
4	Theology of Virtual Reality	49
5	Virtual Reality Mission Field? Discipleship? Church? *Check.*	59
6	The Cultural Impact of Virtual Reality on Physical Church	74
7	The Challenge to Mental Health	80
8	How to Get Started	88
9	Remember, Virtual Reality Is Only the Beginning	92
	Endnotes	95
	About the Author	99

INTRO

Church, Meet the Metaverse

LET'S GET STRAIGHT to the point.

The metaverse needs the Church.

That's not all.

The Church needs the metaverse.

Some will think this is a controversial take. Some will stop reading right here and move on. However, embracing the idea now will give the metaverse Church an opportunity to reach people our buildings are not reaching. So, set aside any predispositions or biases. Maybe even set aside any notions or assumptions about "church" that could be influenced by extra-biblical opinions. The Church, traditionally, is decades behind modern technology. Maybe this is finally the opportunity for the Church to get ahead of the curve. So, look past the shock value here, and let's start to unpack the potential impact.

The metaverse needs the Church. Through this book we will define the metaverse. Be aware that the metaverse, like many things in this post-COVID-19 society, is fluid and constantly changing. What the metaverse looks like will probably change. What is *not* likely to change is Satan's grasp on these metaversal realms—unless the Church intervenes. This is a fact. The metaverse is filled with dark spaces that the Church should not be running away from. Instead, we should be running toward the dark spaces with the light of Christ. *The metaverse needs the Church.* As controversial an idea as a church in virtual reality is, this statement may be the more conservative of the two. Most people may not agree with the ecclesiology of a church in virtual reality, but they recognize the opportunity of the metaverse mission field. So, the Church being involved missionally in the metaverse (for some) is not that radical of a take.

The Church needs the metaverse, on the other hand, is typically where most heads will explode. As we'll show, there's a lot of bad theology centered around the metaverse. On top of that, ecclesiology typically reacts to cultural shifts. The Church is never on the first wave of cultural shifts, instead taking years or even decades to respond. This is why the Church has struggled historically to keep up with technological and cultural shifts. The truth is that these shifts will start to come more quickly and more steadily than ever before. Ultimately, the expansion of metaverse technology will bring about a societal shift. Today's Church has a choice: either lean into these shifts or fight against them. Embracing these metaversal shifts could in fact bring the modern Church into closer alignment with the Church of Acts. That's right—cultural shifts because of modern technology could cause today's Church to align toward a more biblical standard . . . but today's Church would need to pivot.

As you'll find out in this book, the metaverse is not quite here yet. Nor is it going away. But the sooner we embrace the metaverse, the sooner we'll understand what it's capable of—and the sooner we can use its power for Kingdom purposes.

Can we describe the entire metaverse in one book? Probably not. But we can start the conversation. Leadership Network has been at the forefront of developing the metaverse Church, specifically exploring what's happening in virtual reality and beyond.

So, welcome to the conversation! But before we get too controversial, let's establish some common ground with a few basic definitions. First: What exactly is the metaverse?

1

What Exactly Is the Metaverse?

Before We Go Forward, Let's Go Backward

SIMPLY PUT, THE METAVERSE is a series of modern technologies currently being utilized and, in some situations, yet to be released. These technologies, often falling under the umbrella of "Web 3.0," include virtual reality, augmented reality, "mixed" reality, blockchain, cryptocurrencies, and more. Before going too deep into this, let's back up and review some context. Before we can get to Web 3.0, let's understand Web 1.0 and Web 2.0.

Web 1.0: Back When Information Was Enough

About the time Y2K was supposed to destroy the economy as we know it, Web 1.0 was in its heyday. Otherwise known as the "World Wide Web" or "The Information Age," Web 1.0 lasted from 1989 to 2005. Remember all those AOL CDs you'd get in magazines, or going to AskJeeves to search? The amount of information stored on the internet was incredible, and (at least as a kid going through school in the Web 1.0 era) the information was useful. People would go to their Netscape browser and find unrealized information like never before. Seriously, has anyone bought an encyclopedia set or atlas since the birth of the internet? But as technology expanded, culture moved past Web 1.0, embracing new parts of the technology. Content wasn't enough. We wanted community.

Web 2.0: When Standardization Was a Good Thing

Around 2004, new tech companies like Facebook were gaining popularity, and with them came the dawn of social media and ultimately online community. Web 2.0 brought with it new technologies like computer audio and video, which made it easier for people to converse online. Instead of paying ridiculous long-distance rates to a phone company, we started utilizing Web 2.0 technologies to communicate with each other. Shortly thereafter, our phones stopped being merely talking devices and became typing devices. Because they were easier to use, Web 2.0 technologies saw an increase in adoption rates over their 1.0 predecessors.

What's interesting about these new technologies is that they started impacting culture. For example, Web 2.0 adopted terms like *standardization* in hopes of keeping the interaction with technology simple and clean. Twitter would only let you make micro-posts of 140 characters or less. Instagram required a photo to post. Remember Vine? Users had six seconds to communicate on a video clip. Web 2.0 impacted culture in important ways, bringing not only community but also the idea of content standardization through these little devices in our pockets called smartphones. Eventually, even the perception of *people* became "standardized" as we made assumptions based on simplified, stereotypical expressions of a person's race, age, etc.

Web 2.0, in its infancy, brought trust to culture. Think back to 2010 or even 2015. We trusted online sources. We valued people online. That can't be said anymore, which is a source of contention when it comes to Web 3.0.

Web 3.0: The Age of Experiences (Most Likely)

What is Web 3.0? History will eventually tell it best, but from our seat in 2022 Web 3.0 is a meeting of the Age of Experiences and a decentralized worldview.

- Web 1.0 brought us text on a digital page.
- Web 2.0 brought us video in community.

- Web 3.0 will bring us . . . environments? worlds? universes? We just don't know yet.

Defining the Metaverse

It's important to recognize that, at this point, the metaverse is not yet digital. It is a third thing. In very rare instances today would a smartphone be considered a tool of the metaverse. Social media is not yet metaverse, although corporations like Facebook (aka Meta) are working hard to close this gap. We'll talk through specifics, but the heart of the matter is that the metaverse will be the gel that eventually connects websites, the digital community, the metaversal worlds, the economy, relationships, even home entertainment. This is one reason the Church cannot hide from the metaverse. Not to sound "New Agey," but eventually everything will become metaverse. The metaverse will ultimately influence everything.

Let's break down some of the most prominent examples of Web 3.0 technologies, commonly referred to as the "metaverse."

Virtual Reality

The most accepted of current metaverse technologies is virtual reality (VR), which has actually been around since 1960.[1] Referenced in films by sci-fi legends from Stephen King to Steven Spielberg, most people (78% of Americans[2]) have a casual awareness of what virtual reality is. Virtual worlds, designed by people, exist in virtual reality; people wearing virtual reality headsets explore these worlds. VR is different from a video game because, for the most part, it requires wearing a VR headset, giving the wearer an immersive experience unlike anything ever seen before. Virtual reality also tends to be different from just watching a movie, as watching your typical movie is nowhere near as immersive as being placed in the virtual world.

As obnoxious as wearing the VR headset is, as of 2020 an estimated 19% of Americans have tried virtual reality.[3] Approximately 52.1 million Americans (171 million users worldwide) use virtual reality monthly,[4] and an estimated 23 million American jobs will exist only in virtual reality by 2030.[5] VR is not going anywhere, although the real deterrent to the technology is the equipment. Few people like the way they look wearing the VR headset, and the fact that the technology separates the user from physical reality is a detriment to its widespread adoption. Still, there are passionate people utilizing virtual reality regularly for, among other things, community.

While virtual reality offers immense potential and is the primary focus of metaverse technology within this book, VR is just one of several game-changing metaversal technologies currently available or being developed. Let's go ahead and give a brief synopsis of some other metaverse technologies.

Augmented Reality

A perceived limitation of virtual reality is its separation from the physical reality. Augmented reality offers an opportunity to supplement physical reality with graphical overlay. For example, imagine using Google Maps while driving your car, but instead of looking at your phone to see the next turn on an animated vector map, you're looking at the literal asphalt in front of you, where a yellow arrow is telling you to turn right. Driving directions and other pertinent information is mapped out in front of your eyes, either through the windshield of your car or even the eyeglasses you're wearing.

This Heads-Up Display (HUD) can take many different forms but has yet to be released by any major player in the market. The closest thing was Google Glass (circa 2014).[6] At the time of this book's writing, no augmented reality solution has been made available, although Google, Apple, and Meta are all expected to release the technology soon.

The opportunities are endless. Imagine you're walking down Broadway in New York City, looking for a show to watch. As you look (through your augmented reality eyeglasses) at each theater, you see a poster image of each show, including access to bios of the theatrical stars, critical reviews, cost of tickets, available seats, and more. You complete the entire transaction through an app on your glasses.

Sounds a lot like the way you'd interact with your smartphone, right? Exactly! In many ways, features of the smartphone will be integrated into augmented reality. Read text messages or emails through smart glasses. Even know who's calling, answer calls, and talk to people on the phone through augmented reality (using audio transferred through bone conductivity functionality built into the glasses). You're the only one who will hear the other end of the call!

Truthfully, who needs smartphones once smart glass becomes available? Rumors are running amuck, of course, but word on the street is that Apple's smart glasses will replace the iPhone within 10 years.[7] With 1 billion iPhone users currently in the world,[8] it appears that Apple is planning on selling a *lot* of people on augmented reality.

It's important to remember that augmented reality is not virtual reality. Augmented reality is physical reality, augmented . . . or added to. Virtual reality opens immense opportunities to create new worlds, never seen or explored. Augmented reality makes our physical world better. There are philosophical differences between virtual and augmented reality, but if the two can work together, there is immense potential. This brings us to "mixed" reality.

Mixed Reality

Where virtual reality creates vast realities that don't exist physically, and augmented reality projects user interfaces onto reality, "mixed" reality is the combination of both functionalities. Mixed reality puts virtual objects in physical reality to be experienced in real time by multiple people.

For example, Apple has patented[9] technology for its rumored mixed reality headset that will basically invalidate the physical television. Wearers of the purported Apple Glass will be able to look at a blank wall and watch any television show or movie through the AppleTV app installed on an Apple Glass device. Who needs a television when you can wear a pair of glasses? Noise in the room is nil, as the Apple Glass will project the audio into your skull via bone conductivity functionality.

Apple's rumored mixed reality approach, called SharePlay, is (as you might guess) a shared experience as well. If multiple people want to watch the same movie in the same room, the glasses will map the video to multiple eyeglass devices (in sync, of course) to provide a shared experience. Ironically, the only thing heard in the room would likely be the laughter of the people watching! Apple's technology is rumored to also handle long-distance connection. Two people on opposite sides of the planet could be watching the same TV show and talking to each other through a FaceTime audio connection.

Mixed reality will strive to combine virtual objects, or even virtual worlds, with the physical world. Applications are endless. Imagine sitting at a Starbucks on a Zoom call with a remote coworker two time zones away. But instead of seeing the person on a laptop screen, she is sitting in the chair on the opposite side of the bistro table as you view the hologram version of her through mixed reality glasses. More than user interface, these virtual elements will begin to blur the lines between realities. Is the coworker really there physically? Many of the limitations of current technology will obviously start to blur as well. Will Zoom fatigue exist in the same way with a mixed reality version of a person?

Blockchain, Cryptocurrencies, DAOs, and NFTs

There's a metaversal revolution brewing, and it has nothing to do with conversation surrounding these new realities. Instead, there's a potential shift of power, moving away from the "establishment" having it all. Instead, the metaverse is working to empower individuals.

Simply put, "Blockchain is a shared, immutable ledger that facilitates the process of recording transactions and tracking assets in a business network. An asset can be tangible (a house, car, cash, land) or intangible (intellectual property, patents, copyrights, branding). Virtually anything of value can be tracked and traded on a blockchain network, reducing risk and cutting costs for all involved."[10]

Ironically, this definition came from IBM, one of the largest tech corporations in the world.[11] The revolutionary power of blockchain will move to empower individuals, not corporations. Individuals now have access to what corporations had pre-metaverse, and with cryptocurrencies funding decentralized autonomous organizations (DAOs) under guidance of non-fungible tokens (NFTs) and other smart contracts, the Church of 2030 could look very different than it does today. The metaversal technologies of blockchain, crypto, and DAOs open incredible opportunities for a decentralized model of the Church to operate: one that's grounded in a microbusiness structure with the financial potential for doing good in the community.

Blockchain/cryptocurrencies are built upon the principles of transparency and accountability. There is no hiding as these technologies become more and more prevalent. Blockchain and crypto are already bringing about a societal shift through organizational transparency. What does this mean for the Church? Stay tuned.

(Leadership Network is a prominent voice in how churches are utilizing blockchain and cryptocurrencies for Kingdom purposes. For more information on blockchain, crypto, or even virtual/augmented/mixed reality, check out http://leadnet.org/metaversenext).

Metaverse vs. Meta

New technologies will continue to evolve as they are adopted. As we've described, there is a very distinct line between virtual reality and augmented reality. The growth of mixed reality (which is really the blending of the two) will establish how these separate technologies will work together. Consider a decentralized mindset resourced by blockchain or a DAO funded with cryptocurrencies, like Ethereum, and you can begin to see how these isolated technologies will begin to work together. Churches will have an opportunity to blend these technologies.

The lines will blur even more as corporations like Meta (Facebook) meld the space between Web 2.0 (social media) and Web 3.0 (metaverse). As multiple corporations create mixed reality, tools like Apple's rumored augmented reality glasses will continue to complicate things. Suppose you walk into a museum and put on your augmented reality glasses. On the physical wall is an extremely high-resolution projection of Leonardo da Vinci's *Mona Lisa*. Since the physical painting was not in the room, did you actually "see" the painting? Philosophers will be arguing that question for years to come, but the takeaway for the Church is that even interactions in physical reality will be impacted by the metaverse.

The truth is, we will not be able to silo this metaverse technology. Rather than separating physical worlds from virtual ones, the metaverse will blend them. These technologies will not be isolated. They will continue to tie into and build on each other. Web 2.0 and Web 3.0 will become one thing. So, an organizational stance of "We will not utilize blockchain" or "We don't believe in virtual reality" will be problematic in the future. This blended technology will become everything.

The Biblical Meta: "Transcendence" or "Death"?

Yes, that's exactly what *meta* means in the Bible. The Greek (New Testament language) for *meta* loosely translates as "transcendent, among, or after."[12] Meta becomes the glue that takes things to a higher level. It connects everything. It becomes everything.

In fairness, we should probably document that *meta*, in Hebrew, means "dead."[13] It would be interesting to know whether Mark Zuckerberg realized this when he announced the shift of the company name from Facebook to Meta. I'm sure the Church will debate this term in years to come.

The Big, Metaversal Picture

As much fun as these technologies are and will be, they may not be as important as the cultural shifts that will result. The metaverse is literally built on wording like *trustless* and *permissionless*. Crypto exists because people do not trust banks, corporations, and governments. Blockchain empowers individuals, giving them the authority of large organizations and supporting them with the contractual obligations tied into NFTs and DAOs. We're approaching a new era in which individuals will be responsible for choosing their own adventures, not being limited by the vision of the Church. In many ways, today's model of the Church could limit what individuals can do.

The "why" of the Church is shifting because our culture is shifting. To be effective, church strategies must shift as well. As these metaversal technologies solidify, the cultural pendulum will shift more and more toward adoption of these technologies, which will bring into play the idea of decentralization. What does a decentralized Church look like in a metaversal culture? That question is, unfortunately, for another book, but don't fret. Leadership Network is helping you keep up with the conversation in real time. Through Leadership Network's Metaverse Church NEXT we'll continue to explore Web3 technologies, discuss the crossroads where they intersect, and discover how the Church can benefit.

In the meantime, as part of an ongoing series, this book will narrowly focus on metaverse technology and cultural shifts surrounding virtual reality. We'll be publishing more books on other topics soon. Keep up to date at http://leadnet.org/metaversenext.

Virtual Reality and the Metaverse Church Primer: Words Are Important

Throughout this book we are obviously talking about new ideas and new concepts. Right from the start, let's be very clear about what some of these words mean. Words may mean different things to different people in different church environments, so for the sake of innovation within the book please read/infer the following definitions moving forward:

Church Online

Churches that are broadcasting their services online are executing a "church online" strategy. The physical church services are implemented digitally, whether through YouTube Stream, Facebook Live, or the Church Online platform (from Life.Church). Inherently, there's nothing wrong with broadcasting church services online, but their effectiveness in disciple-making is limited compared to other options.

Digital Discipleship

Many churches post-COVID-19 are looking beyond broadcasting church services and are moving their discipleship processes to digital platforms. This includes anything a church does beyond broadcasting its weekend services. What this looks like may vary from church to church, depending on a church's discipleship strategy. As we'll discuss, digital tools will allow scalability and decentralization of your church's discipleship process, enabling you to disciple different types of people in different spaces.

Digital Church

Digital church is any ministry where the primary function (if not the *entire* function) does not require the building. This includes any efforts churches make to reach, engage, disciple, and empower people using cloud-based tools of the Web 2.0 variety. In this book, any reference to "digital" is meant to reference digital tools (mostly 2.0 in mindset) including social media, streamers, websites, etc.

Metaverse Church

Metaverse church is also any ministry where the primary function (if not the *entire* function) does not require the building. This includes any efforts churches make to reach, engage, disciple, and empower people using metaverse tools of the Web 3.0 variety. In this book, any reference to "metaverse" is meant to reference technology from Web 3.0, including virtual reality.

2

Virtual Reality 101

What Virtual Reality Is Not

THERE ARE LOTS OF MISCONCEPTIONS centered around the idea of virtual reality and the metaverse. Ironically, there seems to be even more controversy surrounding the idea of virtual worlds. Here's just some of the negativity:

- **"All new technology comes from fallen angels. It's in the Bible!"** No, I didn't know that. So, the Gutenberg Press was from fallen angels? The radio and television that Billy Graham used so well? Roman roads were from Satan?
- **"Virtual reality will be the birthplace of the antichrist!"** Sounds like a great movie from HBO Max, but I'm not seeing the biblical context here.
- **"Everyone knows the metaverse is of Satan because God didn't make it. Humans made it. It's authentically not God."** Interesting. So, let's talk about the biblical temple (made by humans) and all these church buildings open on Sunday mornings at 9:00 a.m. . . .

I shouldn't make fun. Forgive me. If nothing else, people are passionate about this topic. I would argue that, although quite popular, their passions are misaimed. Virtual reality is different, but there's a misnomer—it may be virtual, but that doesn't make it any less real.

For decades, maybe even centuries, the Church has been afraid of what we do not know or understand. Right now, we're seeing, in real time, a new opportunity unfolding for today's Church. But if we're not careful, we'll miss the very real harvest of the virtual reality mission field, the authenticity of disciple-making in virtual space, and the need for churches in virtual reality spaces.

The Opportunities of Virtual Reality for the Church

The Church has an opportunity *to explore strange new worlds, to seek out new life and new civilizations, to go where no man has gone before.*[14] Pardon the Star Trek reference, but the Church does have an opportunity to be light in areas that do not know of Jesus.

Churches exploring virtual reality report engaging with 80–85% atheists or agnostics.[15] People who want to ask spiritual questions but who aren't comfortable walking into a physical building are putting on VR headsets and walking into churches in virtual reality.

We're also seeing a large percentage of de-churched people doing the same thing.[16] Interestingly, most people who become de-churched are walking away from the physical expression of church, not walking away from God. They've not been burned or hurt by any part of the Trinity, per se. Most likely, they've been burned by people or experienced organizational trauma. The safety and anonymity of the VR goggles allows some people to re-engage with a church community, a church family. We need to remember that *doing* something different *reaches* someone different.

Let's start to unpack what some of these virtual reality worlds look like and hear from people who are already doing ministry in these spaces.

Virtual Reality Communities

AltspaceVR

Owned by Microsoft, AltspaceVR is the "safe" world. It's commonly called "virtual reality on training wheels." Active participants in the community range from 18 to 85 years old and are a solid mix of men and women. And, thanks to Microsoft's involvement, there is a strong level of parental control. You're not going to see naked avatars in AltspaceVR.

Most churches that launch in virtual reality start in AltspaceVR, for every reason already listed. It's safe. (Safe isn't bad by the way. It's a great place to start!) And the best news? Users don't even need a VR headset to visit AltspaceVR. Users can go to http://altvr.com and download software that allows Macs and PCs to have access.

> Imagine, service is going, the worship is blasting, people are worshiping, and as you are walking through the halls of the building, you come across a 20-something-year-old male who is doing the same thing, exploring the building. You walk up to him to greet him and see if there is anything you can help with. He looks back at you and says, "I'm just checking this place out." "Great!" you say. "Let me know if there is anything I can help with or questions I can answer." He then responds with, "You know, I don't really believe in all this stuff, Christianity, that is. I've always had questions but have been too afraid to ask, or afraid I'll offend someone with my questions."
>
> A scenario like that is almost a dream conversation come true. Many of us got into ministry because we have a desire to know Jesus and make Jesus known. But it hardly ever happens that someone walks into the physical building of the church for the first time and openly admits their doubts and spews off their questions about Christianity and, more importantly, Jesus. However, in AltspaceVR, this happens on a regular basis. Virtual reality is where the doubters and skeptics are bringing their questions and finding Jesus.
>
> **—Stuart McPherson, VR Campus Pastor, Lakeland Church**

RecRoom

RecRoom typically has a bit of a younger audience, based on popularity of RecRoom in XBox VR and PlayStation VR. So, RecRoom is likely to be the world where you get cussed out by a 14-year-old. But chances are, he's not telling anyone he's 14; so, there's that. RecRoom gives users dorm rooms, and it's not uncommon to find paintball fights happening.

Here's the story of a user named Alice in the Palace. Formerly known as Alice, Queen of Hell, Alice was a self-proclaimed satanist before she went in to "troll" several churches in virtual reality. Through relationships with multiple Christians in virtual reality, Alice did eventually accept Christ and change her name. Now, Alice serves as a sort of Person of Peace in RecRoom, helping multiple churches solidify their ministries in this virtual world.

> I like RecRoom because the creative community is really cool and [because of] the diversity of the people here. RecRoom is really diverse, creative, forward-thinking, and innovative.
>
> Churches in RecRoom are like hanging out with your friends. It's very casual in RecRoom. You can come and go as you please. There's no pressure here. You listen to someone preach, and after, it's very open in terms of discussions.
>
> My closest relationships in RecRoom were actually outside of the churches. I brought my friends into the churches. Most people in RecRoom are pretty broken. Lots of toxicity all around. Most people just want a friend. You really don't need to push any beliefs on them or anything like that. When you get close to people, they see you and what events you're attending. When hard times come, I find that they bring up the question of God to you because they know you're with the church.
>
> One of my friends says to me, "God must hate me." This actually brings up conversations. You can call that ministry, but that's just being there for your buddy . . . having those conversations. That's the bigger part of how I use RecRoom. There is high value

> in the church services and events, but most of the bonding and conversations come outside of that.
>
> **—Alice in the Palace, RecRoom Person of Peace**

VRchat

Commonly referred to as the "Red-Light District" or the "Wild, Wild West," anything goes in VRchat. There are little to no parental controls in VRchat, so it's very common to see barely dressed, even naked, avatars in this space. It's also been "guesstimated" that 40–50% of users in VRchat are having virtual sex in private rooms. Virtual strip clubs are a thing, and virtual bars are a million-dollar business in VRchat.

That being said, relationships are very important in VRchat communities. HBO recently produced a documentary, *We Met in Virtual Reality*,[17] detailing several relationships (even including a couple who got married) entirely within VRchat.

In general, people in VRchat are very conversational (thus the "chat" in VRchat). Some of the best ministry conversations can happen in this space because people are very raw, real, and open in VRchat. However, the darkness in VRchat (as we'll discuss in later chapters) proves that ministering in this space is not for everyone.

> VRchat is a place where people go to try and not feel so lonely anymore. Just about everyone on the platform struggles with depression, talking a friend out of suicide is a normality. This struggle with depression leads to a vast number of "trolls," which are people that come solely to disrupt.
>
> When you join a VRchat world, expect to hear the worst thing that you have ever heard coming out of a sexualized avatar that's twerking. This is where most churches stop. But when you dig into the "why," you find hurt, you find pain, you find that the reason they are saying and doing these things is for attention because they need to get off their chest everything they have been struggling with that's brought them to the depression they are in.

> VRchat has so many worlds where people try and find a glimpse of hope. From strip clubs, where strippers make a six-figure income, to drinking worlds, where people spend every night—many times with a friend who's known to pass out in his own vomit with his headset still on. They are beggars trying to find the bread. They are thirsty [people] looking to never thirst again.
>
> When you fully understand this, you will be able to sift through the nonsense and be graced in seeing the power of the gospel overcome such extreme darkness. Trust me; I was one of them.
>
> **—Stewart Freeman, Director of VRchat,**
> **Cornerstone Church VR**

MetaHorizons

MetaHorizons (created by Meta, the company formerly known as Facebook) hopes to be the game-changing community of virtual reality. With desires to connect Facebook's social media community to virtual reality, MetaHorizons has huge aspirations. It's important to realize that this is the youngest of all the VR communities.

MetaHorizons has the potential to be an incredible world. It is investing in a virtual community, a musical venue environment, a workplace environment, and other worlds. MetaHorizons currently has the most funding of any of the worlds. MetaHorizons has the potential to be THE community of the future—if Meta/Facebook doesn't screw it up.

Some churches are investing in MetaHorizons today, and their investments are paying off.

> The Horizon community and experience has been incredible. When we started trying to figure out church in VR, we really gravitated toward Facebook and Horizon because of how much we believed in their vision and commitment to being successful. From day one, the entire Meta team and Horizon Facebook/Discord groups for the community have been incredibly supportive

and collaborative. I feel like every week we have someone reaching out who wants to help the church, while the Sunday service experience is still a work in progress with the platform still being incredibly new and not being able to directly put video in our created church yet.

Right now, we stream in Horizon Venue and do the rest of our church activities at our church in Horizon World. The small-group atmosphere and environment has really taken off in our first year and has been the most successful thing we are doing! Our current weekend service audience is about 65% male, and the average age for weekend service attendees is 27, which is really good considering that is a demographic (and often, gender) that most of the Church world struggles to reach. We also are seeing our Alpha Small Groups comprised of a higher participant [percentage] of non-believers compared to the physical church's Alpha Small Groups: about 57% non-believers in VR to 43% non-believers in physical locations.

—Jason Burgoyne, Online Pastor, Sun Valley Community Christian Church

BigScreen

More like Netflix in a VR movie theater, BigScreen is becoming more and more popular for churches producing quality video for their services. Essentially, anyone can easily create a movie theater in virtual reality and broadcast any video, including physical church services, into BigScreen.

This approach, for church environments, does provide opportunities and challenges. BigScreen is by far the easiest world for a church to get into. It's also the least "community-based" of any of these major virtual worlds.

Imagine hosting a private 3D movie event in your own classic theater with a 50-foot BigScreen! With BigScreen, your VR headset allows you to watch TV and movies and play PC games with up to 14 of your closest friends. Rent the movies right in BigScreen or connect directly to your Windows desktop to stream content.

BigScreen environments include a choice of several theaters, a penthouse apartment, a campfire, or even a classroom setting. Customize your avatar, eat virtual popcorn, and even throw tomatoes. With easy access to your PC desktop for slides, videos, and websites and with the cap of 15 people per room, BigScreen lends itself well to a microchurch/small-group ministry. The remote desktop feature is one of the nicest of the VR platforms.

As for the culture of the people in BigScreen, the moderation of users is not as tight as in some other apps like AltspaceVR or Horizons, which means there are more trolls (people who find joy in disrupting events). The room owner does have the ability to kick out and report people who are disturbing their events, and often must do so. Once we get past the crude language and the trolls, however, we have had some amazing and God-honoring discussions in BigScreen.

—Nathan Schindler, Pastor, Fox River VR Church

Others

These are the most popular worlds at the time of writing. All of this can change in an instant—and that's not a bad thing. Now that you know, though, what are you waiting for? Grab a headset (or your mouse) and jump into these worlds. Talk with people. Prayer-walk through these environments and see where God leads you.

We need to understand that these communities are quite literally "strange new worlds," as alluded to earlier. Some are well-policed, safe areas. Some, as described, are very dark and quite chaotic. Officially, there are not yet stats concerning depression and social virtual reality,[18] but unofficially there certainly feels like there is a higher tendency of depression and suicidal tendencies among people in these dark spaces. This is something we need to understand to effectively do ministry in these areas. Today's Church must not run from these dark spaces; we must run *to* them.

The Darkness of Virtual Reality

Let's acknowledge up front that these areas are dark. But we cannot stop at the edge of the darkness just because we're uncomfortable or unsure. We must go in. We must engage. Realizing the bias in virtual reality toward unhealthy attitudes, how can we as pastors and leaders address this darkness in virtual reality? If we're honest, do we as pastors and leaders know how to address this darkness even in *physical* reality?

Virtual reality is filled with a lot of dark places, and unfortunately the darkness can impact some people negatively. Secondhand trauma is a real thing, and we're seeing early signs of it in some active virtual reality churches. What does the VR Armor of God look like? How do we protect ourselves from being overcome or overwhelmed by the darkness? How can we effectively pull people out of the dark state they're in without being pulled in ourselves?

The complexity of mental health has evolved greatly in our post-COVID-19 society, and the Church has far to go in understanding how to address this. Digital Church Network is championing the idea of addressing mental health in the metaverse and digital spaces.We're going to dig deep into these ideas later, but look to http://fam.digitalchurch.network for fresh insights on how we are training and equipping people to minister in these dark spaces.

Haters Gonna Hate. Who Cares?

When we look at virtual reality, the Church needs to see today's opportunities. Much of the metaverse is theoretical, or at least not-yet-released. Augmented reality has not yet been marketed by a major player, so adoption rates are low. Cryptocurrencies are very common yet not quite trusted in some spaces. While it's tempting to qualify virtual reality as tomorrow's opportunity, the truth is that it's quite tangible today.

Church, we have an opportunity to get ahead of the technological curve, for once. Historically, the Church has struggled with adopting new technologies. We fear what we don't understand, and virtual reality is a prime example of something we usually push back on. To quote the theological giant Hellen Keller, "The heresy of one age becomes the orthodoxy of the next."[19]

When Heresy Becomes Orthodoxy

It's interesting that we see this pattern historically. The Church often fights against cultural shifts, claiming them to be heresy. At least in modern history, the Church rarely wins the struggle, eventually adopting the "heretical" view. The heresy of today will become the orthodoxy of tomorrow.

Examples? Church online in the season of COVID-19, for one. Most churches in America pre-COVID-19 thought that broadcasting church services online was a distraction and possibly a violation of God's physical law for his Church. In December 2019, a lead pastor in southern Florida told me that I should confess my sins before God and repent for helping churches broadcast services. In his opinion, I was separating people from God's physical Church. About four months later, essentially week two of the pandemic outbreak in North America, I received a phone call from the same lead pastor begging me to help his church get its services online in the middle of the pandemic. His view of church online went from heretical to borderline orthodoxy in about four months—under the right circumstances.

Gutenberg the Heretic

In his day, some considered Johannes Gutenberg a heretic for his work on the printing press[20] and the eventual printing of the Bible. Putting Scriptures in the hands of everyday people devalued the work of the priest and was said to be very disruptive to the established Church in its time. It's true!

Gutenberg invented the printing press somewhere around 1440. More than 50 years later, Joannis Trithemius wrote *"De laude scriptorum manualium"* ("In Praise of Scribes") in the era of the printing press. His scathing report explains why people of the time should abandon the printing press. Ironically, the document was printed on said printing presses and distributed to the masses. Fittingly, "In Praise of Scribes" is out of print today.[21]

Hundreds of years later, we recognize Gutenberg's work as the catalytic invention that started a movement.

Did you know that Gutenberg's printing press was the prerequisite for Martin Luther's Reformation?[22] Gutenberg's press produced Luther's "95 Theses,"[23] which were distributed around neighboring towns and cities. In many ways, the technology of the printing press allowed Luther's ideas to multiply and spread at an unheard-of rate. Without Gutenberg's printing press, would Martin Luther's Reformation have had the same impact? Victor Hugo, in *The Hunchback of Notre Dame*,[24] references this: "Before printing, the Reformation would just have been a schism; printing made it a revolution. Take away the printing press, and heresy is enervated [weakened, destroyed]. Be it fate or providence, Gutenberg was Luther's precursor."

The Heresy of Virtual Reality

Just as Gutenberg's printing press went from heresy to revolutionary to orthodoxy, we too are poised for the next disruption: virtual reality. Many Church leaders today would view most Web3 technologies as heresy. This lack of vision hurts. Sadly, current Church leaders may not have the best perspective on the metaverse.

Apple, Microsoft, Meta (formerly Facebook), Google, Nvidia, Unity, Shopify, Roblox, Qualcomm, Disney, Adobe, and Nike are all spending billions on virtual reality—and the metaverse overall. And while these corporations are building their tech, the Church is at a crossroads with a decision: do we continue to look at these Web3 technologies as heresy or recognize, like Luther did, that this technological leap is in fact the beginning of something different? Just as the printing press was Luther's precursor, virtual reality can bring a necessary cultural disruption in our churches.

Was Helen Keller right? Will today's heresy become tomorrow's orthodoxy? Time will tell, but we're going to have fun in the meantime!

Innovators Gonna Innovate. Let's Go!

It's important to realize that these ideas are unproven. There *is* some validity in the words of the naysayers. But Church, what would happen if we positioned ourselves not in a place of fearing the unknown but in the seat of learning, understanding, even educating? What if we innovated here?

As we open ourselves up to new ideas, as we understand these opportunities, we will of course see that virtual reality is a lot like physical church. In many ways, it's also different. These differences allow us to reach different people. So, let's start to unpack some of the tangible opportunities and limitations in the metaverse Church.

The Opinions and Challenges of Ministry in Virtual Reality

Relationships

- **OPINION:** Physical church leaders often struggle with the validity of relationships in virtual reality. Virtual reality is essentially a video game. We can't really know what these people are going through. Besides, virtual reality cannot impact the real world. Everything about virtual reality is fake.
- **REALITY:** It may seem unlikely, but churches and ministries doing metaverse ministry are reporting back a high level of relationships, somewhat deeper than even physical relationships. This seems counterintuitive to most Church leaders, who see physical relationships as a necessity. There is validity to physical relationships, but Church leaders need to recognize that relationships, even at the disciple-making level, can

happen in digital and metaverse spaces. There is qualitative evidence to affirm this assertion, and even in our post-COVID-19 society more and more relationships are being developed in the metaverse.

- **CHALLENGE:** The challenge of a church in the metaverse is to truly look at these avatars not as pixels or cartoons but as human people. We need to see that there is a genuine reality in these virtual worlds. With that in mind, we need to pray for the Spirit's leading, prompting us to see these worlds, these people, through Jesus' eyes.

Anonymity

- **OPINION:** Physical church leaders often struggle with the anonymity of virtual reality. We're not looking at a real-life person but a cartoony avatar. We don't know who these people are; we only know a made-up name. We can't minister to these people unless we really know who they are!
- **REALITY:** As virtual reality ministry leaders will attest, not knowing the identity of the people in virtual reality is in fact a strength, not a weakness. As we've learned from digital ministry, people can be very transparent, open, and honest when they're not in the same physical proximity as the church leaders they're talking to. So, it's very common to have deep conversations quickly, *because of* the lack of proximity.

  This is amplified even more by the anonymity of virtual reality. Because of the avatars and the lack of public naming, conversations tend go deep faster in virtual reality than they do face to face. The people/avatars talking rarely have anything to hide because the church leader doesn't know who they are. In a move that may not make sense to those who are not used to VR culture, avatars let people be who they really are—without pretense of being judged by others. This is a strength of virtual reality. People may be at their most authentic state when in avatar mode.

- **CHALLENGE:** The challenge of a church in the metaverse is for avatars to remain anonymous. Keep the mystery in play because often, once the mystery is resolved, conversations shut down, as people assume that the church leader will judge.

Shared Experiences

- **OPINION:** Physical church leaders often struggle with the validity of digital and metaverse experiences, alluding that these experiences are fake/not real. "The problem with virtual reality is that you cannot break bread together. You can't eat together, which is where Jesus was the most real."
- **REALITY:** Let's skip the preconceived notions on avatars eating VR food. At the heart of a physical meal together, or sitting down to a coffee at Starbucks, you and your associate are sharing a physical experience. There are virtual reality worlds consisting of benches around a campfire, where all users do is sit around the virtual fire and talk. You can immerse yourself in hundreds of virtual reality games that have built-in community. In virtual reality, there are many, many ways to share experiences with people you know—and people you don't.
- **CHALLENGE:** We'll talk about shared experiences later in the book, but the biggest challenge facing the metaverse Church is accepting the culture of the metaverse by engaging in these shared experiences. For the metaverse Church to be effective, we need to be doing more than Sunday morning services in virtual reality. We need to be creating and participating in shared experiences.

As we're now starting to get a better understanding of the strengths and weaknesses of virtual reality, let's start to explore how some churches are operating there.

3

The Church in Virtual Reality

ONE OF THE MORE FASCINATING OBSERVATIONS about the metaverse Church is how controversial it is, even among people who are doing, or should I say being, the Church in virtual reality. Many of them are torn on what to call things. Is it a church? Are they pastors—or more like missionaries? Can a virtual reality church be the Church for someone, independent of a physical church? Or do churches in virtual reality complement what's happening as individuals attend both physical and virtual expressions of church?

At the time of this writing, we have fewer facts than theories. And that's not bad. People are innovating, experimenting, and really pioneering in virtual reality. And as culture continues to shift more toward decentralization (we'll talk more about this with the blockchain and cryptocurrencies book coming soon from Leadership Network), there will not be "a" singular model for churches in virtual reality but many different methods to "be" the Church in virtual reality.

Let's dig in and explore in detail what (as of this book's writing) some churches look like in virtual reality.

Examples of Churches in Virtual Reality

Cornerstone VR

One of the first, if not *the* first, church to do physical church as well as church in virtual reality, Cornerstone has had success in their early pioneering. Lead Pastor Jason Poling, an elder, and some key volunteers who were interested in VR and its potential for gospel ministry, started meeting together in 2019 to explore the possibilities and learn more about the culture of the people in VR. With no real experience in virtual reality (jokingly, he didn't even know how to spell the word *VR*), Jason bought a VR headset and started exploring virtual reality worlds. He went from discovery mode to launch in about nine months and has seen Cornerstone VR extend his physical church's ministry well beyond its location in Yuba City, California.

Cornerstone Yuba City is a smallish church, reaching close to 300 people in the northern part of the state. The VR campuses have extended across the country and around the world. In the early days of AltspaceVR, Jason estimated that about 70% of people coming into his virtual reality church were "de-churched."[25] People who had not set foot in a church building in decades were taking their first step back to the Bride of Christ via virtual reality. (Numbers have since skewed from primarily de-churched to more un-churched participants as Cornerstone has expanded into more un-churched environments like VRChat and RecRoom. Still, Cornerstone VR is reaching people who aren't being reached by the church's physical ministry.)

Notice that this sub-300-person church in Northern California is having success in virtual reality. The perception of many churches is that metaverse ministry is really for bigger churches or that smaller churches don't have the capacity to experiment. Cornerstone is proof that this logic is flawed and that smaller churches have potential to do mighty things.

Cornerstone VR started in Altspace (remember, it's the "training wheels" of virtual reality). After having success in Altspace, they extended their metaverse ministry into VRchat, and as of this writing are building into RecRoom. Cornerstone Yuba City has seen the reach of their ministry extended via virtual reality. In fact, they recently hired a pastor to lead their virtual reality ministry. Michael Uzdavines, aka "Goose" in the metaverse, lives in Ft. Lauderdale, Florida, and works for Cornerstone VR remotely.

> Ministry in the metaverse has given me some of the greatest moments of my life as a Christian and as a pastor. When we first entered VR, we did encounter a lot of individuals who considered themselves "de-churched" because of the negative views they had of the Church. I was glad we could help them begin to restore a better image of the Body of Christ through their regular engagement with us. However, something even more exciting began to happen. I began to have more regular opportunities to engage in deep, gospel conversations than I had ever had in my entire life. I had never met so many atheists, agnostics, satanists, wiccans, nones, Muslims, and other non-Christians from all over the world gathering in one place to do one thing: talk about the deep questions of life! I felt like I was entering the Areopagus every day with the Apostle Paul and getting to share the hope of Christ with a constant stream of confused and hurting people. To say the least, it was rapturous!
>
> But it gets even better! The Word of God is truly powerful and won't return void! We began to see people come to Christ! A satanist from the UK. A none from North Carolina. An atheist from Estonia. An agnostic from Finland. And so many more! The harvest was ripe, just like Jesus had promised! We just needed to go to the new field where he was now working: the metaverse.
>
> By design, Cornerstone is a "hybrid" church. We are a local and global (online) church that seeks to help metaverse

members see the value of physical fellowship and local members see the value of online fellowship. We truly believe that as new Christians in the metaverse see the beauty of the Bride of Christ through their fellowship with Cornerstone VR, they will also desire to connect with a physical church fellowship in their local community. Our ultimate goal is to disciple the Christians at all our campuses to grow more in love with Christ and his Church, and to be on mission to share the gospel in all the communities they inhabit, both local and global.

I can't say this strongly enough: it has been pure joy to preach the gospel and disciple new Christians in the metaverse and in Yuba City! But as I said before, I am even more excited about going Home to experience the new reality Jesus has prepared for me and my local and global forever family!

—Jason Poling, Lead Pastor, Cornerstone Yuba City/VR

VR Church

As part of the first church in virtual reality, DJ Soto had a storied career as a physical pastor. Quoted as saying he "never once had an atheist come to [his physical] church to listen to him preach," DJ sees an estimated 80–85% atheists or agnostics in his church services.[26] DJ's church was the first to adopt a decentralized structure in virtual reality. Occasionally referenced as "Bishop Soto," DJ has launched multiple expressions of VR Church in AltspaceVR, time-shifted for access via multiple continents. He also was the first to launch multiple expressions of church into various worlds such as VRchat and RecRoom.

DJ is an example of a pioneer for churches in this space. In addition to virtual reality, VR Church has expanded recently into VR/MMO Church. Massively Multiplayer Online Games (MMO Games) are an example of video games that don't feature a big bad villain to beat (necessarily) as much as they feature a community people can engage in. Games like Final Fantasy, Rust, or Black Desert Online provide community, and with these video-game communities come opportunities to do—to *be*—church in a video game.

> In 2016, my wife and I left the church where we pastored to plant physical churches. About the same time, the Oculus Rift came out . . . the first consumer-available virtual reality headset with a robust app ecosystem. We purchased an Oculus Rift for games and movies, but little did we realize that we would experience one of the very first social VR metaverses called AltspaceVR. When I first entered this metaverse, I was blown away. You think you have seen everything, but I've never experienced a visceral, immersive virtual world where I could talk to people from all over the world that felt very similar to the physical world. It wasn't too long until I thought, "Let's try having church services!" In June of 2016, we launched the VR Church Experience. Five people came to my first church service, which for a physical church-planter would be devastating to have so few people attend, but I was very excited.
>
> The first visitor was an atheist from Denmark who mentioned that even though they don't believe in God, they were curious about church in VR. That was a lightbulb moment to realize that as a physical church pastor, I never had this experience. And that first year of VR Church was mostly people who would not identify with any type of faith. Another lightbulb moment came about a year later when we realized that we were no different than any type of church that existed in the physical world. That physiological shift laid the foundation for a church planting movement in the metaverse by our team, and it strongly influenced others to get into the space after they visited our metaverse church.
>
> From our six years of church planting and church ministry in the metaverse, I've come to realize that the future of the Church is the metaverse. This doesn't mean that physical gatherings or

> locations will go away; it just means that your central ministry hub will be based in the metaverse, from which a whole host of physical and digital experiences will evolve. These are exciting times to live in [and] to experience innovation like this.
>
> **—DJ Soto, Bishop, VR Church**

Calvary Chapel Ft. Lauderdale

A southern Florida megachurch, Calvary Chapel Ft. Lauderdale has taken a unique but effective approach. A video-savvy church with high standards of video production, Calvary Chapel Ft. Lauderdale has been doing virtual reality church for several years now. Instead of going into an immersive community like VRchat, or even AltspaceVR, Calvary chose to host virtual reality church services in the world of BigScreen.[27]

BigScreen is not as extensive a "community" as some of the other VR worlds, but it does allow individuals to watch video-based programming, including church services, in virtual reality. Calvary has used this extensively, broadcasting church services and Alpha courses into Calvary's BigScreen theater. Calvary is even exploring doing Christ-centered recovery via BigScreen.

Whether it's a strength or a weakness (arguably both?), there really is no "target audience" within BigScreen, or at least one has not developed as of this writing. BigScreen, like AltspaceVR, is finding an excellent cross-sampling of all people and has worked very well for Calvary Chapel Ft. Lauderdale.

> We consider our entrance into the metaverse to be a hybrid experience similar to that of venues. Calling it an extension of our real-world sanctuary, we livestream our services into VR via BigScreen. [This includes] a staff of volunteer greeters and admins with the aim of visitors feeling welcomed, loved, and part of the family.
>
> We strongly believe that outside of proclaiming the redemptive value of the gospel, love is our greatest asset as believers.

We simply love the members of the VR community unconditionally and consistently, resulting in their acceptance of us and willingness to listen.

We officially launched in January 2021, and to date we have witnessed over 250 decisions for Christ in less than 20 months.

Approximately 10% return weekly, another 25% bi-weekly, and the rest come in when their VR game-times synchronize with any of our services. Much of our communication with attendees comes via direct chat outside of VR.

At this time, we are hosting 7 different events per week:

- 3 weekly live-stream church services
- "The Chosen" TV series
- Bible Project Studies
- Recovery Ministry VR
- Family Fun-Time (hosted in RecRoom).

It is our hope that other ministries begin to explore reaching the lost in the metaverse, either via their own efforts or by joining efforts already in progress.

For more information about what we do and or becoming a part of Calvary VR, please visit us at https://calvaryftl.org/campus/virtual-reality/.

—Marcos Robert Duran, VR Coordinator,
Calvary Chapel Ft. Lauderdale

Living Room Church VR

Operating more like a microchurch community in virtual reality, Living Room Church creates shared experiences to engage with people in physical and virtual environments. The antithesis of a large church with video production, Living Room Church creates conversational, relational opportunities in physical venues and now virtual reality.

The virtual reality Living Room campus launched during COVID-19, providing opportunities for relationship-building and conversational teaching to help people better understand the Word of God. Living

Room Church VR is another example of doing something different to reach a different type of person.

> Living Room Church VR is a community created to support and celebrate people's spiritual growth. This is a safe zone to connect, discuss real issues, and heal with other imperfect people. We reflect on Scripture, worship, learn, and pray together. Our primary gathering takes place in AltspaceVR every Monday night at 8 p.m. EST.
>
> **—Michael Beck, Pastor, Living Room Church VR**

Choose Your Own Adventure, Virtual Realty Church Edition

As you can see, there are lots of options and opportunities for churches in virtual reality. The wisdom here is not to copy another church merely because they're doing church in a certain way, but to work through some of the nuances of virtual reality and start to develop your own church's unique approach to virtual reality. The ecclesiology of church in virtual reality will take years to figure out, especially as the technology matures. So, don't get lost in that conversation. Instead, start to ask some of these questions:

Church Service? Or Something Organic?

We're finding churches effectively creating church service expressions in virtual reality. We're also seeing more organic/relational approaches being effective as well, such as VR street evangelism, missional opportunities like launching church in a VR bar, or microchurch strategies like Living Room Church. Which is most effective for your church? Just because you do church a certain way physically does not mean you shouldn't experiment and diversify in virtual reality. Maybe there's an opportunity to empower some of your leaders to do something different.

Community Building? Discipleship?

The churches that seem to be effective today in virtual reality are not just doing Sunday morning services; they are building some sort of community for discipleship during the week. Typically, churches are using asynchronous community platforms like Discord or Mighty Networks. We'll talk more about this later, but an important question to ask at this point is what does discipleship mean at your church? What does it look like in a physical space? And how are you going to implement that digitally/in the metaverse?

Don't let evangelism training fall by the wayside. Figure out how to train your leaders to share Jesus in the metaverse! Ask anyone in virtual reality ministry about their time in these worlds, and they will tell you that "the harvest is plentiful, but the workers are few" (Matthew 9:37). In our physical church when someone comes to know Jesus, we connect them to some sort of discipleship community. If your church is not doing discipleship digitally (or in virtual reality) to connect these new Christians, at least take the time to connect them to a physical church in their area or to another church in virtual reality that is offering digital discipleship.

I had a pastor once talk to me about the missional opportunities of his sermons. In his words, he wanted to see people around the world find Jesus through his sermons. He wanted to see people getting baptized on other continents. His missional vision was dynamic. I asked: "What happens after that person accepts Christ?" His response: "Don't know. That's not my problem."

We cannot treat the missional opportunities in virtual reality like drive-thru evangelism. We must have something more for the people we bring to Christ.

Jay Kranda, Online Campus Pastor at Saddleback Church, uses the terms *nearby* and *anywhere*.[28] When someone finds Christ in virtual reality (or digitally) the Church has the responsibility to help that person find an ongoing biblical community to connect to and be discipled in, either 1) "nearby," where the person is physically, or 2) in a digital or metaverse community that can meet "anywhere." Really there's another opportunity of 3) offering the individual both options and letting them decide.

If we're looking to truly engage with anyone in virtual reality spaces, we need to be connecting them into community, whether your church's community or someone else's community. As we'll talk about later, *without community, how we win people is how we'll lose people*.

Edgar Dale's Cone of Experience

The terrifying reality for pastors is that in many ways virtual reality can be better for our ministry, specifically our sermons, than physical reality can be. Proof in point, let's talk about Edgar Dale.[29]

Born in 1900, Edgar Dale was an American educator. After earning his PhD from the University of Chicago, he spent some time working for the Eastman Kodak Company. Dale was a part of some of the earliest documented research on how a film affected people as they watched the projection on the silver screen. Expanding his research, Dale spent most of his life as a professor at OSU, where he created the Cone of Experience in 1946.

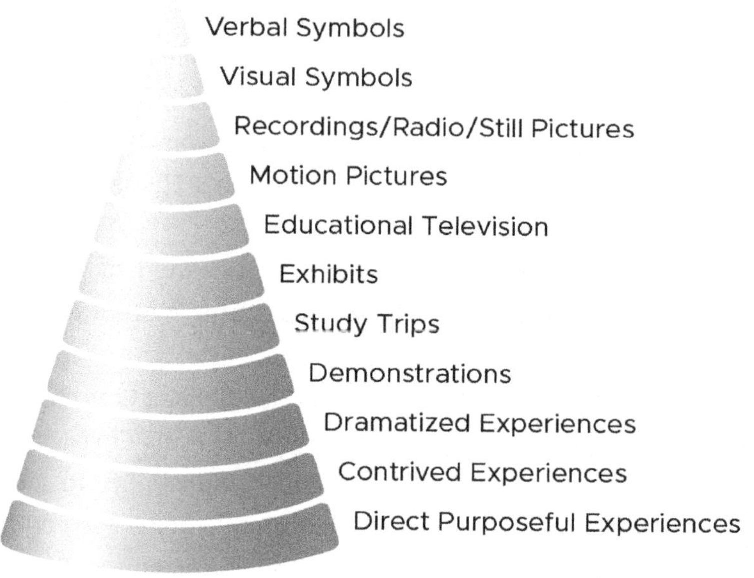

Sermons and the Cone of Experience

Ask yourself this question: how memorable are the Sunday morning experiences at the church you attend physically? Let's defer to Edgar Dale's Cone of Experience.

Most churches today are operating in the top half of the triangle. We're reading books. We're reading articles. We're listening to sermons in person and online.

Here's the hard truth: according to Edgar Dale, a small percentage of people will remember a sermon. Don't believe me? Ask someone in your church what the sermon was about last week. Two weeks ago? Did it stick? In our unscientific research, we have found that people seem to forget Sunday's sermon by Tuesday afternoon.

The bottom half of the triangle is where things get interesting. People forget what they read, hear, or see. These are passive methods of learning. But the bottom half of the triangle (where we see a much more significant percentage of people remembering) involves activity. Very few people remember what they hear, but the majority recall what they experience. For a lasting impact, we should shift toward a participatory approach to preaching that's more conversational, relational, even to the point of experiential.

The audience isn't just an audience.

They're participants.

They're permitted to talk back.

To process it.

To experience it.

To do it.

How do we provide all that in a building? I'm not sure. But let's talk about how that's happening in virtual reality.

Video Teaching or Avatar Teaching?

If you are doing a more structured service, you have all sorts of options. Cornerstone VR has an avatar preaching a sermon every week, while churches like Calvary Chapel Ft. Lauderdale lean on video playback.

Both can be effective ways to communicate a sermon. We have seen increased effectiveness in avatar preaching, especially as we look at the creativity around the sermon and worldbuilding, which is what churches like Oasis Church VR are exploring.

Implementing Immersive Biblical Experiences in Virtual Reality

The leader in immersive church experience at this point is the team over at Oasis Church VR, a new church plant that exists entirely in VR. VRTiger and his wife, Midnight Ad, are bi-vocational co-pastors of Oasis Church.

I recently stumbled into Oasis Church VR and was blown away at their immersive worldbuilding. Their virtual reality services happen in an open-air theater unlike anything I've ever seen in physical space. VRTiger preached his sermon while flying in front of what appears to be a 90-foot screen. And as immersive as that was, it wasn't the best part. When the sermon was over, we went into another world where we experienced the hill of Golgotha from multiple perspectives. Here, the sermon became interactive, participatory, and almost conversational.

Oasis Church's worldbuilding team is incredible. VRTiger and HeEnables, along with volunteers from Oasis and other churches, are active in building immersive worlds each week that reinforce the sermon being preached. Again, very few people will remember what they hear in a sermon, while the vast majority will remember what they experience. Virtual reality is a fantastic opportunity to create experiences, which is probably why Web3 technology is said to be ushering in the "Age of Experiences."

(Ironically, I do not know the actual names of VRTiger and HeEnables. Nor do they matter. Their authenticity shows through, even if they are only known as VRTiger and HeEnables.)

> We desire to create an environment where people can experience God like never before. We want to facilitate a creative,

non-traditional church experience. We are not doing church as our forefathers did.

Each week we curate a custom worldbuild that will be an immersive experience through the Bible. We start in Oasis Church VR with announcements and worship (flight enabled, of course). But the real journey happens when we teleport into the worldbuild outside the church; now, we are missionaries in the metaverse. We dive into a visual interpretation of the Scriptures and scenes with full-on 3D characters, terrain, sound fx, Scriptures, worship moments, and more. The service is now more interactive, as the message is intertwined like a tour through the biblical world, stopping and teleporting to each scene while those participating read the Scriptures. We become part of the story! Seeing the stories in virtual reality, we identify with the characters, scenery, and narrative much more profoundly.

One of the stories our church has experienced has been the parting of the Red Sea. When you visually walk through this scene, looking at Egypt on one end and the Promised Land on the other, with the sound of waves crashing, it opens your senses to feel what the Israelites felt. We've had people break down in tears in the middle of this scene.

There is only so much you can show visually in an IRL [in real life] church service. Showing animal sacrifices or a headless John the Baptist would not typically go over well in IRL church; but in VR, you can push the boundaries and take the congregation into the scene, stretch their emotions, and make them part of the environment. At Oasis Church VR, don't be surprised if a five-story screen pops out of Mount Sinai, and we all have an off-cuff worship experience overlooking the tents in Exodus. It's limitless.

—VRTiger, Pastor, Oasis Church VR

Metaversal Creative Elements? Praise, Worship, and Prayer

Many of the challenges of musical worship in church online transcend into churches in virtual reality. But it's easy to forget that worship is more than musical. Scripture readings, testimonies, spoken word, dramas—even creative worldbuilding and self-directed paths can create worshipful environments that connect people to God. A Prayer Breeze world is a great example of this, helping people connect to God through prayer, even in virtual reality.

> In the world A Prayer Breeze every Tuesday 8 p.m.–9 p.m. PST, a small group of Christians come willing and wanting to pray for people. This all started when, in early 2020, 20 prayer warriors were equipped with Oculus Go headsets to meet every Tuesday and Thursday to pray. But after the prayer sessions, when we were exploring other worlds, we found strangers were open to receiving prayer, which we enjoyed doing with them! So, we created for these world-hoppers a simple, breezy grass world with eight stations to pray in groups privately next to objects like a campfire, Scriptures, animals, etc. Some used the world as a place to pray and worship God after long workdays, allowing them to set their eyes on God.
>
> We kept praying for strangers there because:
>
> - We kept seeing God answer prayer (which is fun).
> - We kept meeting people from all over the world.
> - The coincidences in timing and type of people we met felt God-ordained and would sometimes keep us up until 3 a.m. praying with so many people.
> - Just offering prayer for people turned into one of the safest evangelist opportunities out there.
>
> Over the course of 18 months of hanging out and praying for visitors every Tuesday and Thursday 8 p.m.–11 p.m. PST:
>
> - 4,000 people joined the world.

- 900 unique visitors, of which about 1/3 wanted prayer.
- 200 people favorited the A Prayer Breeze world to return.
- 2 people got baptized at their local churches.
- 1 person prayed the Sinner's Prayer.

Out of the A Prayer Breeze AltspaceVR world-hoppers:

- 1/3 were nebulously New Age.
- 1/4 were atheist.
- 1/4 were Christian.
- 1/6 were other (chivalry-based, heavily warped Christianity, polytheistic, etc.).

Specifics about the history and technical steps can be found at https://www.metaverseministry.net/wiki/A_Prayer_Breeze.

—Chedders, Person of Peace, AltspaceVR

"For his invisible attributes, namely, his eternal power and divine nature, have been clearly perceived, ever since the creation of the world, in the things that have been made. So they are without excuse" (Romans 1:20, ESV). I'm not sure that Paul had virtual reality worlds in mind when he wrote this, but if we can see the nature of God in human expressions like a painting of a sunset, or a photo of people caring for one another, or even in a Veggie Tales video, then why can't we see God move in virtual reality worlds?

Let's unpack a bit more about how creativity in worldbuilding will help us better communicate in virtual reality words.

In the [VR] World, But Not of the [VR] World

"My prayer is not that you take them out of the world but that you protect them from the evil one" (John 17:15).

There's an important distinction to be discussed. When we're talking about the Church and the mission field, we're talking about building the Church in existing virtual worlds where people are. As of the writing of this book, we've not really seen many churches develop their own standalone virtual reality world or application. This is important because, in many ways, the virtual reality Church has recognized the need to go where people are, as opposed to trying to get people to come to them.

In 2022, there are 50.2 million VR users in the US.[30] That's roughly 15% of the US population. (This percentage is increasing and will increase quickly once augmented reality comes out.) Interestingly, it seems that percentage is not true in churches. In fact, it's rare to find more than a handful of VR users at any given church. What we're finding as we talk with churches doing virtual reality ministry is that the ministry is not about the physical churches connecting with people they already know. Virtual reality ministry is more about connecting with people the Church has yet to meet.

This is the challenge of building your own virtual reality app, as opposed to building in an existing world such as Altspace or VRchat. Yes, building your own independent app *does* give your church more control, and possibly more freedom. But unfortunately, you'll then be tasked with marketing your virtual reality app and trying to draw people to *you*, as opposed to being in more of a public space where you can go to *them*.

Jesus probably didn't have virtual reality at heart when he said this, but the missional charge to remain in the world is strong throughout his teachings. Maybe one day it will make sense for churches to build their own virtual reality apps and become isolated from existing virtual realms, but the missional implications today are huge, and we should take advantage of the opportunities to build in realms like VRchat and AltspaceVR.

For the virtual reality haters and naysayers out there, remember that God had to *force* Jonah to go to Nineveh. But that'll be a conversation for another day.

How We Win People Is How We Lose People

It's very easy to look at the power of virtual reality and geek out, to dream crazy dreams and imagine a high-tech virtual world based on

Web3 technologies. And these dreams are great. But recognize that, without community, how we *win* people is how we *lose* people.

If we depend solely on creating over-the-top experiences in virtual reality and don't focus on community where people can connect, then we risk losing people when someone else creates a better over-the-top experience. As we look at the strengths of existing virtual reality churches, we're seeing that successful churches are creative with the technology, but they're also focused on community.

Comparing Digital and Metaverse Evangelism

Let's switch gears for a second and specifically talk about the Church and digital ministry, as in online church services and social media broadcasts. Ad campaigns and SEO can be an effective way for your church to digitally meet new people. Through Facebook ads or Google searches, your church can connect with people who have never been to your church building. Hopefully your church not only has a worship service people can be involved in, but also an engagement process/discipleship plan that operates in the digital space. The chances of "converting" these new viewers from the online church service into the digital discipleship process is slim—especially here in America. Simply put, your church does not have relational equity with someone who is clicking on a Facebook ad. It is possible to get these new viewers engaged in your church, but the process is difficult—and to be honest, there are far more thorny bushes and rocky paths than fertile soil in this process. Finding the fertile soil is a challenge.

In digital spaces what has been more effective, although rarely used, is when people who are active in their church invite people they know. This type of church visitor, one who is connected through a friend, is more likely to stick around because of relational equity not with the church, but with this friend. The friend is the reason the visitor will stay. The friend increases the chances of the visitor engaging in the discipleship process.

Church, the best digital resource you have right now is sitting in your pews. What if your church equipped its members and attenders to reach their circles of influence—digitally *and* physically? How different could your church look if you gave people a spiritual purpose to use social media?

Let's compare this to evangelism in virtual reality. The challenge here is that your church members and attenders may know a couple people in virtual reality. (Remember that 15% of America uses virtual reality.) So, a very small percentage of people have access to these headsets. This is where event-based evangelism like Alpha has been successful virtually, and even more than that, "street" evangelism that engages people by talking about Christ. Using shared experiences or evangelism (structured or organic, like church in a VR bar), all these methods of connecting with people virtually will only be effective toward disciple-making if we are faithful in building relational equity.

Relational Reformation and *The Other Half of Church*

In recent months, one of the most discussed books within digital ministry circles has been Michel Hendricks and Jim Wilder's *The Other Half of Church: Christian Community, Brain Science, and Overcoming Spiritual Stagnation*.

Throughout the book, Michel, a former spiritual formations pastor, talks about some of the struggles of discipleship in today's Church, paralleling these challenges to brain science. According to the book, the typical Western church is not operating in a way that creates consistent disciples. Publicly, Michel has often talked about a second Reformation of the Church, a "Relational Reformation."[31]

The challenge for relational disciple-making has been invigorating to several digital and metaverse pastors, even though in the book Michel holds a negative view toward digital ministry. Truthfully, I agree with Michel and his perspective on hate-filled social media and on the isolation of watching church services online. So, it is imperative for church in virtual reality to move beyond broadcasts, shows, and services, and to pull people into community.

> Can I have a sense of belonging digitally? Maybe the better question is, can I have a sense of belonging remotely?
>
> As an online minister reading *The Other Half of Church: Christian Community, Brain Science, and Overcoming Spiritual Stagnation* by [Michel Hendricks and] Jim Wilder, I couldn't help but think through how vital a sense of belonging is to one's spiritual growth. It's not by accident that we are born into families and that Jesus did his discipling in the context of a group of friends. But can belonging happen remotely? Seems like it can if you look at the way the Apostle Paul writes.
>
> When I read Romans 16 or Colossians 4, it's more than obvious the sense of belonging that Paul is fostering amongst the communities of faith there, introducing them to each other and networking them like family.
>
> What's more surprising is that when Paul wrote Romans and Colossians, he had never met these people personally yet found a way, from his prison cell, to remotely and deeply connect them into a sense of belonging to a greater remote network of believers. Sounds a lot like remote digital ministry if you ask me. We're just using more modern tools.
>
> If Paul can network people together with a sense of belonging remotely with single-use 2D parchment and pen, how much more easily can that sense of relational networking, belonging, and disciple-making be reached using dynamic video conversations and 3D environments that are much more analogous to the ways people interact with one another relationally?
>
> I suppose that the deep relationships of family and friends, especially when bonded together by the same Spirit, are true regardless of distance.

Just ask Jesus; we've been growing as friends and family for a long time, and we've never met IRL.

—Jason Morris, Product Manager, RESI video streaming

The Church That Runs a VR Bar

Talking about doing something different? "Let's open a virtual reality bar!"

These were the words of Joey Santos, Global Integration Pastor at Christ's Church in Mason, Ohio. A megachurch in the area, Christ's Church has always had an aggressive digital strategy. In a recent metaverse Learning Community through Leadership Network, I challenged Joey to launch something ministry-focused in the metaverse. His idea to start a VR bar wasn't a total surprise to me. I've always respected Joey and his church for their ability to think outside the box. Joey may spend time developing systems digitally, but he spends a lot of time pastoring the local community at one of the church's micro locations at a local bar in town. A self-proclaimed "Pastor of the Bar," Joey is reaching different people by doing something different.

Yes, his church meets in a physical bar in town. Maybe this idea is controversial, although I've met several people lately with similar "bar church" stories. When Joey pitched the idea of launching a bar in the metaverse, it did not come as a surprise. He's been bringing the church to bar people for years.

Bars in virtual reality? Well, they're a little different from physical bars. Virtual bars in worlds like VRchat are a thriving business model, as people pay for memberships and subscriptions for exclusive access to shared experiences. Some of these experiences may be a bit shady, as VRchat is considered by many to be the cesspool of virtual reality. Alcoholism, drug use, prostitution, pornography—I've met people who have been sexually assaulted in virtual reality dark spaces like the back rooms of a VR bar. Where there is darkness, the gospel light shines the brightest. The Church MUST go into VRchat, and the more I listened to my friend Joey talk about opening a virtual reality bar, the more I wanted to be a part of this vision.

This is not a venue for a church service, mind you. The goal is not to open a virtual reality sports bar with lots of video screens where we can broadcast church services on Sunday. The more Joey and I talk about the idea of church in a virtual reality bar, the less we talk about sermons or services. There's nothing wrong with church services in virtual reality. People are getting connected to God through those services. But we're looking and longing to do something different. We're looking to capitalize on building relationships. Some may call it pre-evangelism. The goal is to create an environment where we offer mission-minded people opportunities to share experiences with others and point people to Jesus through these relationships.

I've been working on a theory lately about the metaverse. It applies digitally to social media as well. In the metaverse, relational equity (friendship) is at a premium. So, the "why" for a virtual reality bar in VRchat is simple: we want to connect with people. We want to build relationships with people. We want to create an environment where our people can create shared experiences with the people who come into the bar. It may seem silly, but shooting pool or playing cornhole or merely talking to someone in virtual reality builds relational equity that establishes trust, which gives the listener the ability to "hear our heart" once we do talk about Jesus.

So, we're launching a bar in VRchat? Some of you are excited by the prospect. Some of you are mortified. "Why not do a coffee shop in VRchat?" (I've had this conversation multiple times.) There is some legitimacy to the question. But we see this as an opportunity to create multiple environments for shared experiences. The VR bar is just the beginning.

> The idea to meet people where they are is not new. That's something we all do from time to time in social settings. Usually, we meet without much pre-planning or expectations, but just to

> get together, socialize, and share a few things we may have in common.
>
> That is why we thought about taking a step a faith and bringing our church to a bar. Every weekend we meet regularly at a bar to do church in a very unorthodox way. But we've learned that what makes church at a bar impactful is not the regular gathering as much as our regular presence at the bar, meeting people, talking with people, sharing life. Most people attend Church at the Bar on Sunday because of the connections that were made during the week, and that is only possible because we decided to go and meet people where they are.
>
> That is how we see moving this in-person experience to the VR world—create a place to experience social connection and allow real-life conversations to take place that will take a deeper dive into Bible teachings.
>
> **—Joey Santos, Global Integration Pastor, Christ's Church**

The takeaway here is that to reach the metaverse we need to think outside the box. Church services in virtual reality will reach certain people, but we have opportunities to do new things (like run a bar) and reach new people in that space. Some may think this is sacrilege. Others will think this is a great idea. I'll stand before God one day and ask him. But until then, let's continue to think differently and do different things to reach different people.

Interested in being part of the Church in a VR Bar? If so, check out http://churchinavrbar.com.

4

Theology of Virtual Reality

WHAT IS THE THEOLOGY of virtual reality? First and foremost, we need to realize that virtual reality is a tool. We often discuss tools like virtual reality, or even social media or other digital services, as though they are agnostic.

It is important to note that at this point there are very few people speaking into the theology of the virtual reality church. Many are making blanket negative statements without examining the opportunities and the challenges in the metaverse. As we've said, the ecclesiology of the metaverse will take time to develop as the technology matures (allowing for more connectivity).

In the meantime, Jason Poling (Lead Pastor of Cornerstone Yuba City/VR) is working on an informed perspective on the theology of virtual reality. If you're interested in his holistic take, swing over to http://cornerstoneyc.com/metaversetheology. We won't take time here to write a complete review of the theology. Instead, we'll address some of the key points surrounding theological issues of church in virtual reality.

Technological Agnosticism: The Unbiased Role of Tech

It's important to acknowledge that technology such as virtual reality is unbiased theologically. It is neither Christian nor satanic. Like just about anything, it can be used by people for good or for evil. Right now, everything from tarot card readings to virtual sex is happening in virtual reality, so it is safe to say that the enemy is utilizing the technology quite well. Yet there is an opportunity for God's people to utilize this for his glory. Titus 1:15 reinforces, "To the pure, all things are pure, but to those who are corrupted and do not believe, nothing is pure. In fact, both their minds and consciences are corrupted."

Does the Bible say that virtual reality is evil? Evidently not, if "all things are pure." Obviously, Scriptures do not embrace virtual reality specifically. Theologically, we can only look to the philosophies that Scripture gives us and apply them to the virtual realm. As we apply these Bible-informed theologies to virtual reality, let us examine how much of our reactions are based on extra-biblical inconsistencies like opinions and biases.

To the Ends of the (Virtual) Earth

The early Church clearly understood their charge. Jesus did not mince words: "But you will receive power when the Holy Spirit comes on you; and you will be my witnesses in Jerusalem, and in all Judea and Samaria, and to the ends of the earth" (Acts 1:8).

Jesus did not specify AltspaceVR or VRchat. His descriptions of Jerusalem, Judea, and Samaria were indicative of the city, country, and region where he was. He then used a global designation: the "ends of the earth." One can easily interpret these words to mean physical locales—your city, your country, your region, your world. I interpret them to also mean virtual reality worlds like RecRoom and MetaHorizons. These (virtual) worlds exist within the physical world, so it is not hard to stretch "the ends of the earth" to include them.

Here's an interesting theological quandary (for another day). Once space exploration solidifies, and we see colonization on other planets like Mars or even the moon, do we have a biblical mandate to take the gospel, to be his witnesses, beyond the ends of our earth and to other planets? (There's your next book, Leadership Network! You can thank me later.)

Have Humans Finally Created a Box God Cannot Enter?

Many people have said that Christians should not have a place in virtual reality because it is: 1) manmade, 2) not of God, or 3) inherently evil. Let's examine these ideas at a deeper level. If virtual reality is inherently evil and not of God, does this mean that humans have finally built a place where we can hide from God?

The idea of hiding from God goes all the way back to Adam and Eve in the Garden. Genesis 3:8-10 tells us that "The man and his wife heard the sound of the Lord God as he was walking in the garden in the cool of the day, and they hid from the Lord God among the trees of the garden. But the Lord God called to the man, 'Where are you?' He answered, 'I heard you in the garden, and I was afraid because I was naked; so I hid.'"

Examining Scripture, we know that Adam and Eve were not truly hidden from God. Jeremiah 23:24 (ESV) speaks to man's attempt to hide, from God's point of view. "Can a man hide himself in secret places so that I cannot see him? declares the Lord. Do I not fill heaven and earth? declares the Lord."

The idea that we can hide from God is not biblical in any way. To think that humans have created virtual reality as a vacuum in which God does not exist is an atrocious misunderstanding of God. "And there is no creature hidden from His sight, but all things are open and laid bare to the eyes of Him to whom we must answer" (Hebrews 4:13, NASB).

Is This Church? Parachurch? Mission Field?

Is what we're doing in virtual reality an ecclesiologically stable church? Is virtual reality a missiological opportunity for evangelism that needs to feed into physical buildings for discipleship and community? Can biblical community exist in digital and metaverse spaces? What's funny is that even the innovators, those experimenting with ministry in virtual reality, do not agree on answers to these questions.

Some think virtual reality can be an ecclesiologically stable expression of biblical community. Others do not.

The biggest arguments for and against churches in virtual reality use the same Scripture as justification: "And let us consider how we may spur one another on toward love and good deeds, not giving up meeting together, as some are in the habit of doing, but encouraging one another—and all the more as you see the Day approaching" (Hebrews 10:24-25).

On a personal note, I do believe that we can be a biblical community, that we can "spur one another on toward love and good deeds" digitally, and that by meeting in virtual reality we are "not giving up meeting together, as some are in the habit of doing." Some people are comfortable in these digital and metaverse communities and scoff at the idea that we cannot be the Church in these spaces, while other people's heads are practically exploding at the mere suggestion of not meeting in a building on Sunday morning.

Who's right? Let's ask God when we get to Heaven; but for now, let's not limit what is happening in virtual reality by trying to force a label on it.

Does Virtual Go Physical? The Challenge of Embodiment

Along the same lines of "What exactly *is* this thing?" we can ask, "What happens next?" The missiological implications of virtual reality are more commonly accepted. Most people will believe the concept that virtual reality is a mission field, even if they are not sure of the church context. So, we bring someone to Christ; then what?

There are many stories of people who have found Jesus in virtual reality because of individuals and churches "engaging" the mission field. These people embraced the gospel and found Jesus—all while wearing the VR headset. Sometimes their stories turn south at this point, as the new Christian is encouraged to join an "embodied" community but cannot find a welcoming church in a physical space. In one situation, a person attended two different physical churches over three weeks, only to find that "not one person said anything to me [even] once."

People who are comfortable in virtual reality environments may not be comfortable in physical environments. There is an opportunity here to utilize the virtual reality ecclesia to equip people to better embrace missiological opportunities physically. We'll talk more about this in a later chapter, but we commonly refer to this standard as "Online to Offline." Simply put, the gospel that we hear in our online (virtual or digital) world must influence our offline (physical) relationships—otherwise we're merely creating consumers. This principle is built upon James 1:22: "Do not merely listen to the word, and so deceive yourselves. Do what it says."

Virtual Sacraments? Communion and Baptism?

How important are biblical ordinances like communion and baptism to virtual reality? Jason Poling, of Cornerstone Yuba City, states:

> Without baptism and communion, a Christian gathering cannot be called a "church."[32]
>
> **—Jason Poling, Lead Pastor, Cornerstone Yuba City/VR**

Jason goes on to describe how the acts of communion and baptism are re-imagined in virtual reality. More and more, we are seeing the idea of "re-imagining" these biblical ordinances as a common theme among virtual reality pastors.

Baptizing Avatars

Several churches in virtual reality will go so far as to baptize avatars. This may be the most controversial idea in this book.

Let's examine Scripture.

In John 3:5, Jesus so clearly paints the picture: "Very truly I tell you, no one can enter the kingdom of God unless they are born of water and the Spirit." Scripture points out that the act of baptism is important.

That is not in question. But do baptisms in virtual reality count, or is there a spiritual role for physical water?

1 Peter 3:21 reminds us: "This water symbolizes baptism that now saves you also—not the removal of dirt from the body but the pledge of a clear conscience toward God. It saves you by the resurrection of Jesus Christ." Here the author explains that the water molecules are not physically doing the act, but the importance lies in the decision of the individual being baptized to pledge "a clear conscience toward God."

1 Corinthians 12:13 seems to reinforce the symbolic (more than physical) nature of water. It's not the baptism of water, but of the Spirit. "For we were all baptized by one Spirit so as to form one body—whether Jews or Gentiles, slave or free—and we were all given the one Spirit to drink."

There are different ideas surrounding virtual baptism, even among churches in virtual reality. Some will not baptize. Some will baptize, but they challenge the new Christians to be baptized physically as well. Some think that virtual baptisms are standalone and "count" without needing to be reinforced with physical baptisms.

At the heart of the discussion, biblically we see the symbolic act of baptism as the alignment of the Holy Spirit in the life of the person being baptized and as a point of celebration for those witnessing the baptism. The idea of virtual reality baptism, as controversial as it may seem, is not that controversial through the lens of Scripture.

Virtual Communion

Almost as controversial, the idea of virtually partaking in communion (the Lord's Supper) has also stirred the pot among traditional church leaders. Is virtual communion sufficiently respectful to Christ's sacrifice? Does taking communion virtually give us an opportunity to examine ourselves as Scripture calls us to?

Let's start with Paul, as he reflects on the Last Supper in Scripture. First Corinthians 11:23-26 (ESV) says: "For I received from the Lord what I also delivered to you, that the Lord Jesus on the night when he was betrayed took bread, and when he had given thanks, he broke it, and said, 'This is my body, which is for you. Do this in remembrance of me.' In the same way also he took the cup, after supper, saying, 'This cup is the new covenant in my blood. Do this, as often as you drink it, in remembrance of me.' For as often as you eat this bread and drink the cup, you proclaim the Lord's death until he comes."

There is something intimate about eating together—the safety and trust of sitting across the table from a person, looking into their eyes. Yes, communion can be simulated virtually, but just because you *can*, does it mean you *should*?

There are those who believe that the physical elements, the bread and the wine, do literally change into Christ's body. If you subscribe to this belief, then the idea of virtual reality communion is likely not possible, unless you subscribe to the belief that the pixels on the screen can change to Christ's body. It may sound outlandish, but if molecules and atoms of a cracker can transform into the literal body of Christ, then can God not change pixels as well? Talk about faith!

If you simplify what's happening during communion: breaking bread, drinking from the cup, remembering Jesus . . . then yes, these three elements can be done virtually. Christ's sacrifice can be honored by those participating in virtual reality environments. Physical elements can be taken outside the church building by people wearing headsets. This presents a great opportunity to remember Christ's sacrifice in public settings such as homes, campuses, or restaurants—and perhaps this kind of remembrance is even more meaningful than if it took place in a church building or in virtual reality.

Is Virtual Reality Real? The Better Question: Is Even Physical Reality Real?

Hold on. This may get interesting. *What is real?* Is virtual reality real? Is our physical reality real? Is the heavenly realm real? Are we consistent when assigning the label of "reality"? What does the Bible have to say? Jason Poling, Lead Pastor at Cornerstone Church Yuba City/VR has spent some time diving into this. Here are some of his theological insights:[33]

> All this is still a lot to wrap our minds around. Virtual reality is real, even as it copies the real things of our physical reality. This might sound novel, and therefore, somehow wrong. But the idea of alternate realities that are absolutely real, even though they are copies of other realities, is not new. In fact, it is very biblical. This is something for which we have not flexed much mental muscle in our biblical theology. The Book of Hebrews gives us a peek into God's view of reality. In chapters 8–10, the author reveals that the Old Covenant priests were serving a "copy" or "shadow" of the "true" or "real" things that exist in the heavenly realms. The Tabernacle and all its artifacts and rites were mere copies of the heavenly Tabernacle. We are told in Hebrews 10:1 (ESV) that literally all the details of the Old Covenant and its law were "but a shadow of the good things to come instead of the true form of these realities."
>
> This first-century language of shadows and forms would have been well understood by the readers of the Book of Hebrews. This does not mean the author was beholden to the speculative philosophies of Plato and Philo, thinkers who both gave rise to the "shadows and forms" language. It simply means that in the common grace offered by God to believers and unbelievers alike, philosophers like Plato may have uncovered some truths about the nature of reality apart from direct revelation from the Spirit of God. But now, here in Hebrews (as well as in places like Colossians 2:16-17), the Spirit of God directly inspires the author to use this common philosophical and cultural language to unpack reality in a biblical and theological fashion.
>
> We know from Genesis chapter 1 and other Scriptures, and from our humble attempt at inductively establishing a theology

from these sources, that God exists in a reality outside of time and space. This means that the time, space, and reality we live in are all constructed things made by him. They are very real, but that does not make them the ultimate reality. Therefore, we must admit that our experienced reality is derivative. It is a copy of the heavenly reality. But going further, even the heavenly reality itself is derivative in that it too is just a copy of the ideas that existed in the mind of the eternal God. The author of Hebrews famously writes in chapter 11 verse 3 (ESV): "By faith we understand that the universe was created by the word of God, so that what is seen was not made out of things that are visible." The "invisible things" implied in this text appear to refer to the ideas of God, which he fashioned *ex nihilo* ("out of nothing") into physical entities by speaking them into existence.

All this serves the author of Hebrews' overall point. The heavenly reality contains the "true" or "real" things, as Hebrews 8:2; 9:24; and 10:1 suggest, and so we must not get trapped in the very physical, temporal, and ceremonial ideas and practices that are products of the old way in which God covenanted and interacted with his people. Rather, we must see that the New Covenant in Christ has opened us up to a much greater reality, one that is identified with the more realistic forms of the heavenly and spiritual realm than with the "shadow" forms of this present realm. In other words, [the author] is saying these heavenly and spiritual things are the supremely real things, yet still reminding us that at the deepest level, even these things have their existence in the most real thing of all: the mind of God. To be absolutely clear, all of this mind-blowing yet still biblically faithful theology does not mean our reality and the things in it, like the Tabernacle, are fake. It just means that the things in this reality are copies of the things that are ultimately real.

—Jason Poling, Lead Pastor, Cornerstone Yuba City/VR

(Books can—and should—be written solely on the theology of the metaversal Church. This important discussion is, however, for another time. If you're interested in a real conversation on this work, check out "Theology of the Metaverse" by Jason Poling [http://cornerstoneyc.com/metaversetheology], but realize this is merely scratching the surface of the true conversation to come.)

5

Virtual Reality Mission Field? Discipleship? Church? *Check.*

IT'S VERY EASY TO GET LOST in the theology of virtual reality, which—if we're honest—is not yet developed. Ecclesiology will continue to fine-tune itself as new metaversal technology is released. This is why we're using terms like *pioneers* when describing churches going into the metaverse. (Unfortunately, pioneers get slaughtered. Settlers get rich.[34]) We need these innovators to carve the path that the future Church will follow.

While the biblical functions of a church are being reimagined, it's easy for the mainstream Church to be critical of the pioneers. As the great theologian Taylor Swift says, "Haters gonna hate."[35] So while it's easy to be critical of the idea of the Church in virtual reality, let's examine some of the more missional opportunities that virtual reality holds for these pioneers.

The Virtual Reality Mission Field

As we examine the missional implications of virtual reality, we're seeing that virtual reality truly allows us to do something different. And we're not talking about avatars or building your virtual reality church building to look like the Avengers Tower in outer space.

We can reach different people, via different methods, in different environments, and empower a different type of leader.

Reaching Different People

Gamers are not officially a people group, but if they were, video gamers would be one of the largest unreached people groups in America[36] (not that video gamers are exclusive to virtual reality ministry). Churches in VR are also seeing an increase in agnostics, atheists, and the de-churched in these virtual worlds. (And a surprising percentage of these people are gamers!)

> There are many advantages to doing ministry in virtual reality, one of which is just the pure boldness and authenticity that one expresses from behind a headset. In any given event, we at Lakeland VR will see a variety of different people join with a variety of different motives. Sometimes we will have your typical Christian. Someone who loves Jesus, and they are so excited to see churches in VR. You'll also have your investigators. These people join the event really just to be a wallflower and see what's going on. They won't talk to you, or anyone else for that matter, the entire time. It's also not uncommon to see them venture off and just go exploring, which brings us to another type of person in VR, the explorer. These people just hop around from event to event and world to world because they just want to see what the VR world has to offer. They might say a couple of words to you, or they might just zoom right past to begin their exploration. Of course, just like any other online-based platform, you get your trolls. These people are just there to cause a ruckus. They'll cuss you out, make obscene gestures, do ridiculous stuff, say outrageous things, all until you finally give them the boot. So, why do ministry in VR? It seems like the soil is hard. It sounds like the majority of the people have no interest in church. That may be true, but that's also true in real life as well. More and more people are stepping away from the Christian faith every day. So, again, why VR? Well, because of the opportunity to introduce someone to Jesus. After all, that's

what we are commissioned to do as followers. [But] it's also to introduce people to the right Jesus. What do I mean by that? Glad you asked.

One other group of people that you'll find attending your VR church or ministry are those who think they know Jesus, but in reality, they have a misunderstanding of who he is. We had the honor of having a practicing Muslim attend one of our church services in VR, and for the most part he best seemed to be like an investigator. He joined our event and didn't really say anything or go very far from our spawn point. At one point, one of our team members went to greet this individual to welcome them to Lakeland VR. . . . From that, we learned he was Muslim who was just wanting to learn. As we got into conversation, he began by asking what we believed and why we were doing church in VR. He eventually began to ask us questions about God and Jesus. The conversation then turned to us asking him about his beliefs and what he knew about Jesus. It was at that moment we learned that even though he said he knew Jesus, he was never introduced to Jesus—let alone knew Jesus. . . . My heart broke at that moment because I knew he didn't know my Jesus, but I was also excited about the opportunity to clear the waters and introduce him to Jesus. VR provides opportunities like this every single day. Yes, the soil seems hard and thorny at times, but it's also true that the harvest is plentiful. In VR, people are meeting Jesus, sometimes for the first time—and sometimes, they are meeting the right Jesus for the very first time. In VR, you can know Jesus and make Jesus known.

—Stuart McPherson, VR Campus Pastor, Lakeland Church

For whatever reason, the people being reached by churches in virtual reality are not people who are going into physical church buildings. It's a different group of people. For this reason, churches doing virtual reality ministry don't see virtual reality as a threat to their physical building ministries. People aren't saying, "Let's skip church this Sunday, stay at home in our underwear, and go to church in virtual reality instead."[37] The fact is that the perceived threat of virtual reality churches (or digital churches, for that matter) and their impact on physical attendance has yet to be realized by any church! Sure, there are reasons that physical attendance is dwindling. It has very little to do with VR headsets, though. (Dwindling attendance *does* have a lot to do with some of the cultural shifts we're seeing because of the metaverse. We'll talk about this later.)

Via Different Methods

The fact that there is not yet a solid model of virtual reality ministry may be the best reason it is effective. At present, there are too many unique expressions to make one status quo. And remember, the uniqueness allows these churches to reach different audiences. A physical microchurch will reach different people than a smaller, generational church who will reach different people than the megachurch. Similarly, the different methods of a church in virtual reality will allow for different types of people to be reached. The micro approach of Living Room Church VR will connect with a different type of person than the small-church approach of Cornerstone VR, which will be completely different from the megachurch approach of Calvary Chapel Ft. Lauderdale VR. None of these methods are wrong, mind you. All of them are necessary to reach different people.

Realize this as well: the best methods are probably yet to be discovered. We need some pioneers who are willing to get out there and "take a bullet!"

In Different Environments

As culture moves more and more toward decentralization, not only will you see more emphasis on ministry happening outside the building, but you will also see churches move toward adapting multiple strategies—or more specifically, toward helping individuals within their church start different strategies in different environments.

Churches like VR/MMO Church will hopefully become the norm—not because of their lack of a building, but because they are multiplying their church into different virtual reality worlds and different video game environments. Every new world and every new game are mission fields in the making. Have you ever live-streamed a national prayer walk while playing a flight simulator that's virtually flying across America? ("Who do I know near Kansas City? Post some prayer requests. How can I pray for you?") Maybe one day you will.

The Church, for hundreds of years, has limited its definition of "community" to that of physical proximity. This is why the Church struggled in COVID-19 season—because we were literally locked out of the only way we defined it. Whether or not you believe that "Church" ecclesiology functions in these digital and metaversal communities, you have to recognize these virtual communities as a mission field.

Empowering Different Leaders

Often, the leaders we see championing ministry in virtual reality are not the people we see leading in physical ministry. These people are capable of leading but are more comfortable in virtual reality than in physical reality. It seems weird, but sometimes we've even seen these kinds of people even be uncomfortable with a web camera on in Zoom. Whether it stems from identity issues, body issues, or something entirely different, there is a subset of virtual reality where people struggle with who they are. So, empowering them to do ministry in virtual reality is not a negative thing, but a way to build confidence.

All of this reinforces the idea that virtual reality ministry is not in competition with physical church ministry. Virtual reality is just a different mindset.

Discipleship in Virtual Reality

In the post-COVID-19 culture we're currently navigating, do not freak out about the fact that people are not in the church building. Freak out about the fact that people have no spiritual purpose when they're *not* in the church building.

Virtual Reality Discipleship Standard: Online to Offline

There has been a digital church theory coming out of Saddleback Church over the past decade that sets the standard for all digital and metaverse churches, virtual reality included. Simply put, the "Online to Offline"[38] idea states that the gospel we hear in the online world must influence our offline relationships. If it doesn't, then all we are building is consumerism. Our churches must do more than broadcast church services or create virtual reality experiences. The standard is creating disciples, and the goal is for these disciples to exude Christ physically (as well as digitally and virtually).

We've seen digital churches grow and develop into physical micro locations. We've not seen it yet, but what would it look like for a virtual reality church to disciple people to create physical micro locations? Or lead people to the place of discipling family members or coworkers? Can a virtual reality church disciple people to impact their physical circle of influence? Can we be the vehicle that transfers the gospel out of virtual reality into physical reality? Early success stories suggest that we can,[39] but time will tell.

More Than a Weekend

As we look at different opportunities for discipleship in digital and metaverse spaces, we recognize the need for discipleship to be built in and around community. Honestly, that discipleship . . . that community . . . is not grounded in a Sunday morning experience. Shared experiences, including Sunday morning, can be the front door to a discipleship community—but so can any number of things.

Mark Lutz, Lead Pastor at Lux Digital Church, once said that 15% of his time was spent actively involved in something having to do with the weekend service (on a social media platform called Twitch) and the other 85% was spent developing community[40] in a platform called Discord. The challenge of a virtual reality church is to create shared experiences in virtual reality that engage people and encourage them to dive deeper into community.

There is one massive difference between physical and virtual/digital that needs to be discussed. We have seen this trend multiple times. People in the digital/metaverse space connect to community before they connect to Christ. In evangelical church services held physically across the United States, we see how people can make quick decisions for Christ. Moved by the eloquence of the pastor speaking or the power of the chorus being sung, people raise their hands to accept the grace of Christ. In virtual reality (as well as digitally) we also see salvation transpiring like this, but it happens almost as often when people engage through community.

Async 101

Life-changing decisions for Christ, because of the relational equity needed, often take longer digitally and in virtual reality than decisions in physical church. That's good and bad. But we're seeing discipleship happen much more quickly in digital space than in physical. Why?

At the time of this writing, many churches operating in virtual reality are utilizing digital tools like Discord to complement their virtual communities. Let's be up front. We can't wear VR headsets all day, every day—at least the current iteration of headsets. To reinforce community, many virtual reality churches are falling back on asynchronous community tools that allow for community 24/7/365, even without the headset.

The heart behind asynchronous communities (commonly called "async") is that the platform allows people to engage in community asynchronously (outside of time/at any time). Async communities are typically formed around topics or channels, not content. Content exists in the channels. Also, asynchronous approaches work best not when content creators and admins lead, but when the users run the community on their own. When the community is self-sufficient, as we'll discuss, there are immense opportunities to multiply.

Discord is currently one of the more popular asynchronous tools. Other asynchronous platforms include Slack, WhatsApp, Facebook Groups/Messenger, Mighty Networks, and others. Facebook and Instagram posts are similar in that responses are text-based. But Discord is the primary option for virtual reality churches because of the current alignment of virtual reality and video games. As the virtual reality platforms mature and we see growth of augmented and mixed reality, you will see some diversification away from Discord. As of today, though, Discord is by far the most powerful platform. But it is *so* effective in reaching gamers that it's often difficult to get non-gamers to utilize the Discord platform. Utilizing a mix of text-based communication (like WhatsApp) as well as audio and video calls like what you'd find with Zoom, there's a lot of functionality within Discord that community can be built around.

The Collaborative, Multiplicative Opportunity of Asynchronous Discipleship

As discussed, spiritual formation is happening primarily in platforms like Discord. So, the opportunity before us connects shared experiences to discipleship opportunities. We're seeing this in newer church plants like Oasis Church VR, who connects their weekly virtual reality church services to their church's Discord channel. We're seeing more mature virtual reality churches like Cornerstone VR—a church that has developed multiple expressions of their church in VR worlds like Altspace VR, RecRoom, as well as VRchat—utilize a single Discord channel across all their shared experiences. Essentially, the asynchronous community serves as a centralized piece of the metaversal puzzle.

This offers an incredible opportunity for worldbuilders and content creators in both digital and metaverse spaces. Rather than feeling the pressure to create and manage an asynchronous discipleship process, there are opportunities to partner with other churches and individuals. There are some important fundamental shifts at work here, such as:

Decentralization: The metaverse will bring cultural shifts toward decentralization. Governments, corporations, and even large organizations will not be trusted. Through the movement of blockchain/cryptocurrencies and DAOs, we will see smaller, more-focused organizations develop. In many ways it's not as important that our organizations operate as standalone, fully functioning, turnkey providers. Instead, we should recognize that we can work together with others, which leads to the second shift . . .

Collaboration: This is a crucial lesson for the Church, and society overall. Rather than feeling the burden of mastering outside their strengths, churches and individuals can partner up to accomplish combined goals. Could you see virtual reality worldbuilders, streamers, and community builders working together toward a common goal of discipleship? The move toward decentralization allows us to operate independently, but the collaborative environment found in the metaverse allows us to work together toward a common cause—not united under a brand but united by a purpose. We can be friends on mission!

This Is Bigger Than You or Your Brand

I recently visited a popular megachurch on the East Coast of America who is running an experiment. They have created a decentralized community, unbranded/white labeled and separate from the church. The goal of their online community is not to make people aware of the megachurch brand—in fact, there is no mention of the church's name anywhere in the community. The goal of this church's community is digital discipleship. As of the time of this writing, thousands of people have connected in this community. How did these thousands of people find this unbranded community online? Through a single social media influencer who is partnering with the church. This one Christian influencer, who has tens of thousands of followers, successfully connected more than 3,000 people to a digital church.

Imagine what can happen if this church connects two or three more influencers to this community? We would see massive growth in the discipleship community! Now, what happens when three people from the community decide they each want to start their own community to reach a different type of person or to disciple in a different way? What would it look like if the megachurch supported this expansion, not like a digital multi-site location, but allowing for contextualization like a digital church plant would, striving to reach different types of people? Realistically, you could have two different digital churches, reaching different people, empowering numerous streamers or virtual reality environments, operating out of one church. We just multiplied digital churches!

Asynchronous communities provide immense opportunities for collaboration and multiplication, but only if the parties involved remain open-handed and selfless. Metaverse ministry is amplified when we move past our branding and recognize that the potential in working alongside others is greater than doing it alone.

Shared Experiences Are Just the Beginning

Worldbuilders and crafters of virtual experiences are very important to the virtual reality church. Just as Sunday morning service is not the only thing your physical church does, we cannot base the entirety of a church's discipleship on a Sunday morning expression. Driving people to an asynchronous community is the perfect way to continue building disciple-making relationships.

That may be a controversial statement. Can someone develop disciple-making relationships utilizing digital tools like Discord? If you think that statement is controversial, then it probably won't work for you. The truth is, some people can connect, engage, and build relationships digitally. This is based more on their comfort level with technology than anything spiritually. Can God move in virtual reality, or even in platforms such as Discord? Let's hope so, because if humans finally created a device that God cannot enter, then the world is in trouble.

Digital discipleship, holding to the Online to Offline standard, is not an easy task. We're suggesting that we can build relationships with people digitally, and through those relationships and digital tools we can change the way these people act in the physical world. We're not just talking consumer-level discipleship here; the standard our churches should see is one of action.

The Challenge of Consumeristic Discipleship

Digital and metaverse ministry, because of the natural tendencies of isolation, will lean toward consumerism. An individual can watch a church service in isolation. That same individual can be well-read in Scripture and other Christ-centered material. The challenge that virtual reality churches must overcome is to not just focus on virtual reality, but to drive disciples all the way to physical reality.

James 1:22 [paraphrased] says: Do not just be hearers of the word [digitally, or in virtual reality] and so deceive yourselves. Do what it says [in the physical world too].

With the newness of churches in virtual reality comes the opportunity to define how the Church will operate in the space. Looking at the vast mission field available in virtual reality, as well as the scalability of digital discipleship, it is imperative that churches do not settle on knowledge-based discipleship. We must continue to obey the commands of Christ and the calling of the Great Commission to "Go."

Can virtual reality help grow physical reality? Can we see physical micro locations or workplace ministries born out of virtual reality expressions? What needs to happen for virtual reality churches to become the diaspora that creates new ecclesia? This is the true test of discipleship in virtual reality churches. Reaching people in virtual reality is a huge first step, but the long play is not about the people *gathering* but more about them *scattering*—in virtual reality and physically.

Defining the Church in Virtual Reality

I recently sat down with a metaverse missionary who is responsible for discipling hundreds of gamers on Twitch, Discord, and in virtual reality. He told me that "we need a new word for *church*." He went on to say that "when I talk about the Church, people are immediately turned off. Not because of God or Jesus, but because of bad experiences people have had historically with the organization of church." His insights reminded me of the old Mahatma Ghandi quote, "I like your Jesus. I do not like your Christians. They are so unlike Christ."[41]

One of the biggest challenges of the physical church, specifically the megachurch model, is contextualization. A popular trend in recent years has been assimilation, specifically how to "assimilate people" into the Church or the Church's belief system. Let's get people coming on a weekend, attending a group, serving on a weekend, and let's not forget giving. For assimilation to work effectively, we need to normalize the discipleship process. What assimilation misses is contextualization. For the Church to effectively reach people moving forward, we can no longer treat all people as though they are the same. Assimilation is not the answer. Contextualization means recognizing that people have differences and understanding that these nuances are effectively opportunities to connect.

The Challenge of a Million-Person Church

Let's have some fun. Do you think it is better to have a million-person church or 100,000 churches of 10? Is the goal to grow a church as big as possible or to multiply churches like they're rabbits? There are certainly pros and cons to both models.

- A million-person church has influence due to its size, but it is unwieldy. It's not agile. And it's very hard for people to not be treated like a number.
- 100,000 churches of 10? Limited resources financially. Limited staff to support the ministry.

But in metaverse and digital spaces, the idea of 100,000 churches of 10 plays very well for a couple of reasons:

1. Digital and metaverse spaces allow for less-expensive ministry than physical spaces. Limited financial resources are not an issue.

2. Smaller churches can contextualize much easier than larger churches. Digital/metaverse churches can contextualize much easier than physical churches.

Contextualization in Worlds

Currently, churches operating in virtual reality are paying attention to context. Each world operates differently. What works in AltspaceVR won't work in VRchat. RecRoom is completely different contextually than MetaHorizons. Would we have the same approach to church in New York City as we would in Buford, Wyoming? Of course not! The same treatment should apply to virtual reality. Recognize the people in these spaces, and pray about how to contextually connect with them.

The challenge of the Church in virtual reality is not how to execute a one-hour service. It's not how to re-imagine biblical ecclesiology in digital/metaverse spaces. The challenge is recognizing virtual reality as a community and finding that community—those people—worthy of having a church.

Contextualization in Methods

Due to the experimental nature of virtual reality, we're seeing a lot of different methods of church. We've got physical churches doing virtual reality church. We're seeing virtual reality churches planted with no physical buildings. There are megachurches creating virtual reality expressions of the same. There are microchurch expressions in virtual reality, operating as such. We've got group-centric churches in virtual reality. We've got missionaries walking around virtual reality who aren't even sure that what they're doing constitutes as church. Let's not forget about the church in a VR bar! But honestly, we need *each* of these models (and a ton of other crazy ideas) to effectively reach the metaverse.

The challenge of the Church is not to look at virtual reality and see a simple solution. We're not at the place where it's simple copy and paste. (And maybe we never will be!) Whether or not the "playbook" for virtual reality churches ever gets built, we need to understand that our churches must contextualize to reach different people, and the importance of this is magnified in the metaverse.

Don't Die on the "Church" Hill

Is virtual reality really a church? Some say yes. Many say no. Remember, it's very early in virtual reality's ministry lifespan. As technology matures, you'll continue to see the ecclesiology mature as well.

The ecclesiology of the virtual reality church, or even digital expressions of church, cannot really be resolved in 2022. I've talked with biblical scholars who tell me that the ecclesiology of a virtual reality church may take 20 to 30 years to develop. We may not even get a solid answer on this until 2050! Why? Because while God is the same yesterday, today, and forever, our culture changes. Constantly! And as culture changes, so does our strategy. Church has historically been averse to cultural change. Just look at how quickly churches are pushing for "butts in the seats" again, post-COVID-19. This is why the Helen Keller quote mentioned earlier rings true: the heretical struggles of today will be orthodoxy tomorrow. (Churches will struggle to be on the bleeding edge. Ecclesiological conversations are almost always reactionary instead of assuming the best and being proactive.)

So, passionate virtual reality supporters out there on the battlefield of virtual reality ecclesiology—don't die on that hill. That is not today's battle. As fun and frustrating as the metaverse ecclesiology conversation is, the metaverse as a mission field is far more important for today. Tell the stories of life change from the metaverse mission field! Celebrate the disciple-making and sending out that's happening via meta tools and resources! Because, while people can argue ecclesiology all day long, they will never be able to argue what God is doing through you personally and those you're reaching in the metaverse.

Naysayers of the Church in virtual reality—take a moment. Breathe. Redirect the passions you have against the *Church* in the metaverse and focus that intensity on reaching the *people* in the metaverse. Maybe spending time in the metaverse mission field will give you a better understanding of the metaverse ecclesiology. And, Church, give some grace and understanding (and encouragement!) to those people, those missionaries, those churches who are experimenting in the metaverse. Let's pray that they are successful in better understanding the metaverse today so that we can effectively reach the people our buildings are not reaching.

6

The Cultural Impact of Virtual Reality on Physical Church

WE'VE TALKED A LOT HERE about how the Church needs to influence virtual reality and, in large part, the metaverse. Let's turn the tables and discuss briefly how the metaverse will influence the Church.

There are several cultural shifts that are happening because of the metaverse. Web3 technology will influence society, and as society shifts so should the strategy of the Church. This is not a move away from the gospel. Christ doesn't change. But let's look at how corporations and organizations are shifting as a result of the advent of the metaverse.

Recognizing the Purpose, the Place of the Individual

One major cultural shift resulting from the metaverse is the role of the individual compared to the organization. Old-school mindset says that the individual's role is to serve the direction of the Church. Recognizing the decentralized view that's coming, there's an opportunity to focus on the individual. Interestingly, isn't this how Jesus would see the world today—as individuals instead of organizations? In many ways, this changed viewpoint aligns with Christ's—as do other metaversal cultural shifts on the horizon.

From Institution to Individuals

Many churches have been traditionally focused on the Church as an institution. From a practical standpoint, the "butts in the seats" metric, or even the "nickels and noses" metric, leads to this. Pre-COVID-19, the larger the organization, the stronger the perception that the organization was healthy. We did whatever was necessary to keep our institutions healthy.

One of the metaversal shifts we are seeing is the transition from the institution being of highest value to the focus being on individuals. Post-COVID-19, we see now that power is in the hands of individuals, not institutions. When properly equipped, these individuals will take the gospel to places our institutions cannot go.

From Brand to Platform

Because of a focus on the institution as opposed to the individual, the Church has too often been worried about its "brand." In the new era, the Church cannot focus on brand. Our focus must be on people, as it always should have been.

As we look at the future of the Church, we see opportunities to decentralize and to empower individuals in new and unique ways. As our institutions lose influence, the Church must move to give people the necessary platforms in physical, digital, and metaverse spaces to be the Church in ways that our institutions cannot. The Church must move to a position of equipping people, encouraging them, and enabling them to use their voices.

From Controlling to Releasing

If you're not listening to the people around you, then you'll quickly find yourself surrounded by people who have nothing to say. This is the burden of today's churches, who typically have positioned themselves to control situations.

Instead of feeling the burden of responsibility in these situations, the Church must release control and empower others. For the Church to diversify ministry models, we need to recognize that we are not able to know everything about everything. Instead of controlling, we must release. The ability of our churches to multiply in the digital/metaverse space depends on it.

Finding the Posture and Purpose of Change

As the metaverse, through decentralization, puts more emphasis on the individual, we as the Church need to change our posture to do so as well. While our biblical purpose, the gospel, is not changing, our organizational purpose must change.

From Planning to Experimenting

Remember, much of the metaverse is yet to be defined. How we are using this technology is an experiment. The pandemic was a cultural accelerator. As a result, culture adjusted quickly. Gone are the five-year plans and forecasts. Instead, organizations are shifting to a more agile approach.

Shifting to a culture of experimentation causes the organization to admit one thing right off the bat: "We don't know." We *think* something *may* happen, but rather than assuming that to be true, let's run an experiment to see what *will* happen. The experiment nullifies the risk of trying something on a large scale. In fact, a culture of experimentation allows us to do more risky things in controlled bursts.

If we as the Church want to be effective in reaching people post-COVID-19, if we want to reach different people who our building strategies are not reaching, then we *must* experiment with some different things.

From Hierarchies to Networks

For years, the Church operated in hierarchies, taking orders from the top and passing them down. There are several reasons why society is shifting away from hierarchies—moral failures, lack of judgment, and toxicity, just to name a few. These aren't just charges against the Church (although the human part of "the Church" is guilty of much). Moving forward, we must change the way we lead.

Rather than top-down leadership and being close-handed, what does it look like to be collaborative and open-handed? In this new season, top-down leadership is limiting. Rather than feeling like your church needs to be the turnkey solution for everything, shifting toward a collaborative approach like networking will extend your abilities. In this new season, 1+1 equals much more than 2.

From Privacy to Transparency

While we're moving toward trying new things and admitting to ourselves that we don't have all the answers, we need to be transparent with others. Pre-COVID-19, there was a drive for organizations to hold that five-year forecast close to the chest, to not share. There was a hierarchical mindset that put the leader in control and created a need-to-know environment. Shifting away from a hierarchy to a network mentality requires moving away from privacy and toward transparency.

Metaversal shifts will lean toward decentralization, and the opportunity for organizations like yours to multiply is a very realistic one. Society is celebrating organizations for being transparent and open; therefore, private organizations will seem inauthentic to most. This is another reason transparency is so important.

From Strengths to Weaknesses

Post-COVID-19, it is very easy for organizations to operate in their perceived strengths. COVID-19 did a great job of breaking . . . well, everything. Healthy organizations are now forced to operate in their weaknesses. Economic shifts. Supply chain issues. Employee shortages. Travel concerns. At the time of this writing, the world is in a collective mess. But even as things seem broken, we need to recognize the opportunities in the mess.

The metaverse will force organizations into a place of weakness. Virtual reality can be messy, but there are opportunities in the mess. Cryptocurrencies are a perceived risk, but there are opportunities in the risk. The push for decentralization due to blockchain will likely be a struggle for many organizations in the days to come. Ultimately, organizations need to shift culturally away from only operating in their strengths and to recognize the opportunities to work in their weaknesses. Even options like strategically partnering or collaborating with someone who can help overcome those weaknesses is a win.

Rediscovering the Voice of the Church

The metaverse will change the voice of the Church. At least it should. Jesus is the same yesterday, today, and forever. That is not changing. May we always listen to the Spirit who leads. However, because of these cultural shifts, the voice of the Church has the opportunity to shift as well.

From Omnichannel to Contextualized

The decentralization shifts of the metaverse will cause many churches to embrace the idea of contextualization. COVID-19 plays a part in this strategic shift. Pre-COVID-19, the goal of the Church was to get its voice on all social channels. This approach, called "Omnichannel," was focused on exposure and reach rather than contextualization. What we're discovering about Omnichannel (in the church context) is that its weakness is engagement. Sure, we can get eyeballs on our stuff, but unless it's contextualized for the audience, they're not likely to respond. Omnichannel gets churches lots of viewers but provides limited exposure on who they are.

The move away from Omnichannel toward contextualization recognizes that different people behave differently. They look for different things and view life differently. These differences are a huge opportunity for churches to contextualize, to decentralize their mindset. It's not about exposing many people to a single message; it's about crafting a contextualized message that can be spread using decentralized methods.

From Making Statements to Asking Questions

This may be one of the most awkward shifts for churches, the idea of shifting from "talking" to "listening." Ultimately, this shift is grounded in the fact that individuals don't trust organizations. Also, in both digital and metaversal spaces, trust is earned through understanding, which is birthed by listening.

Instead of feeling the urge to control the situation through talking, there's a cultural shift where listening and understanding are paramount. We do see this exemplified through technology-based relationships, but we must realize that this ideology goes well beyond the metaverse and digital reality.

Said a different way, we need to move away from one-direction communication via content and instead focus on two-way communication through shared experiences in community. Using more question marks than periods will take your church a long way in the metaverse.

As you're reading this, you're starting to see how the metaverse will influence the Church—hopefully for the better. Now let's dig into one of the largest challenges the Church will have in the metaverse.

7

The Challenge to Mental Health

I WAS SPEAKING at one of the Exponential Regional Conferences a while back when I met her. In her early twenties, she was attending a breakout session where I was talking about digital ministry and exploring what virtual reality could be for the Church. Specifically, I remember getting to the place in my talk when I was sharing about the virtual reality world called "VRchat." It was then that I noticed her. She was hard to miss.

The first time I said "VRchat," she about fell out of her seat. Truthfully! I just kept talking, thinking she had fallen asleep in my talk. I laughed to myself a little and tried to look in another direction so as not to make her feel awkward. Unfortunately for her, I said "VRchat" about five other times in the next couple minutes, and every time I said the words, she had a visible physical reaction. What was going on?

I wrapped up my session quickly and dismissed for a break. Wondering whether I should have a conversation with the girl, that question was answered when she didn't get up to leave like everyone else. I motioned to a woman in the room, asking her to stick around to help me talk with this girl. I started asking questions of the girl, and she shared openly.

She was a relatively young Christian. She grew up in church but walked away from it as a teenager. Just recently she again found Jesus and became active in reaching people via digital ministry. But then she started talking about her life before Jesus and how she was sexually active, in avatar form, in this world called VRchat. She started telling stories in which she was a willing participant in the experience. She also told stories in which she was manipulated into having sex, even describing several of these experiences as rape.

What she was describing is an unspoken reality of VRchat. I've heard it estimated that 40–50% of VRchat users can be found at any given time having sex in private rooms. VR strip clubs are common. VR bars can net owners as much as six figures a month (in real US dollars!), renting out back rooms to facilitate experiences that would make most people blush. For VR users in these dark spaces, virtual experiences are quite real. And for this young girl, these experiences did not stop once the headset came off. She's carried emotional damage from these virtual experiences into her physical being.

With surprising intensity, she looked me in the eyes and said, "Jeff, because of what happened to me, the Church has no business being in VRchat."

I looked back at her and simply replied, "What happened to you is exactly why the Church *has* to be in VRchat."

Church, Are We Prepared to Minister in Dark Spaces?

Internet addiction is a real thing. In virtual reality there is a greater degree of suicidal tendencies, depression, alcoholism, drugs, pornography. And really, it makes sense. In worlds where there are "no consequences," of course we're going to see addictive behaviors. Church, are we ready to reach the people trapped there? Remember, we cannot run from the darkness; we must run into it.

I know pastors and ministers in these spaces who talk to me about secondhand trauma. Doing ministry in these dark spaces pulls them into the darkness. Constant exposure to addictive behaviors and the darkness therein creates unhealthy spaces that are difficult for some of us to go into. Digital Church Network is leading the way through the darkness with Christ-centered recovery organizations as well as mental health organizations to address many of the mental health concerns that come out of the metaverse. We do recognize that, because of the darkness, metaverse ministry is not for everyone.

The Challenge of Who Should Go

In the early 2000s, I was a twenty-something newlywed when I received a call from my friend who was coming into Miami (where I lived). At the time, he was doing ministry with XXXChurch, an organization working to reach people within the porn industry (among other things). He was visiting Miami Beach to attend the world's largest pornography convention (or something) to do ministry with XXXChurch. Essentially, he and others from his team were going to Miami Beach to walk around the largest porn convention on the planet and share with others about Jesus . . . and my friend was inviting me along.

I'm going to be honest. At that time in my life, I was not strong enough to go into this dark space. Situations in my own life . . . my new marriage . . . it was one of the few moments where I could audibly hear God speak to me: "This is not for you."

My situation and response don't mean the ministry of XXXChurch is wrong or bad. I'm not invalidating their ministry in the slightest. In fact, it's *incredibly* valid. So, I positioned myself in a role where I could support the team going in, even if I could not join them on the journey. There are plenty of instances where people have felt that "this ministry, this opportunity is not for you." But the metaverse ministry, this opportunity—it *is* a calling God has put on the hearts of many. For those who have been waiting for such a time as this, "We do not lose heart" (2 Corinthians 4:1).

Who among you is called to minister in virtual reality? If you are not called, then how can you support those who *are* being called?

Mental Health Obstacles in the Metaverse

Recently I was speaking at an intimate gathering of Christian tech leaders about the ministry opportunities in the metaverse. It's always exciting to share about the current activity, as well as the soon-to-be opportunities in virtual reality and beyond. As I monologued through stories of life change in the metaverse, they were not surprised. VR was new territory for them, but as Ecclesiastes tells us, there's nothing new under the sun. They understood the mission field before us utilizing this new technology. During the Q&A portion, I may have received the best question I've ever been asked about the Church in the metaverse: "Jeff, with everything you're seeing and doing in the metaverse, what gives you pause?"

My response: "The mental health implications of the metaverse are ridiculous. These areas are untested. Virtual reality is filled with addictive centers, and it thrives on a lack of control. Alcoholism, drug use, prostitution, pornography—some of these worlds flaunt vices with little to no consequences for these actions. Many of these worlds do not leave the vices in the virtual world alone, instead pushing action into the physical world as much as in the virtual. Virtual reality in itself is an addictive medium. Yet we are sending pastors and ministers into this space. Are they prepared to handle the dark spaces?"

The Need for Mental Health in the Virtual Reality Church

In virtual reality, we often find a higher percentage of people with mental or emotional trauma. Maybe the isolative tendencies of virtual reality attract damaged people, or perhaps an avatar gives them more freedom to speak their truth. Hard to tell, but through Digital Church Network we've seen far more instances of depression and suicidal situations in metaverse environments than in digital or even physical environments.

Therefore, we must figure out how to overcome mental health obstacles and empower the Church to reach people in virtual reality spaces and beyond.

The Case for Christ-Centered Recovery in Virtual Reality

Pre-COVID-19 there was a very strong bias against the idea of Christ-centered recovery working in digital and metaverse spaces. While Alcoholics Anonymous and other secular agencies were leaning into digital spaces pre-COVID-19, most Christ-centered recovery programs were against the idea of recovery in the digital community, much less in virtual reality.

Well, COVID-19 did what it did, and most Christ-centered recovery programs were forced to engage in digital community. In the same way that the Church must re-imagine its ecclesiology to work in digital or metaverse spaces, Christ-centered recovery organizations may need to rework their practices, or at least re-imagine some of their principles, to be effective in virtual reality.

It's interesting that some recovery organizations thrived in the COVID-19 season and learned lessons for the future. Yes, some recovery organizations are still hesitant to go into digital (as well as virtual reality) communities, but some are making headway. Recovery orgs that are shifting are a prime example of what it takes to: 1) recognize and operate in their weakness, and 2) be comfortable in an experimental culture.

> Because of The Fall, everyone experiences brokenness in life. Pain is the common language of humankind. In the metaverse we find people seeking an alternative reality, searching for escape, pain relief, and shared experiences beyond the physical world. What an opportunity for those of us who have had our eyes opened to spiritual realities in and beyond the physical realm! Regardless of where someone is physically located, we now can meet that person at his [or her] point of pain,

authentically share the healing and freedom we have found in Christ, and join with him [or her] in actively practicing biblical steps of health. This is the heart of Christ-centered recovery and the Great Commission. When has there ever been a better opportunity to reach hurting people from "the ends of the earth" or even those from our own communities who might not dare walk through our church doors?

—Nate Graybill, National Director, re:generation

Re:generation is a biblically based discipleship program offering recovery from any number of struggles. Interestingly, Nate and the re:generation team have seen a higher completion rate of the program when offered online/digitally compared to physical classes. Yes, digital participants are more likely to complete the course than physical ones![42] At the time of writing, re:generation and other Christ-centered recovery groups continue to explore the idea of Christ-centered recovery in virtual reality, learning as they go.

(Look for more on Christ-centered recovery programs in virtual reality as Digital Church Network works with other organizations globally, championing the idea of recovery in the metaverse.)

Armor of God, Virtual Reality Edition

Christ-centered recovery is vital to the success of ministry in virtual reality, but if we aren't careful, recovery can be generalized or discarded as yet another "event" in a church. Based on the sheer volume of people who are carrying some level of trauma in virtual reality, what does it look like to equip pastors, planters, or even everyday people to handle difficult situations in virtual reality? How do you talk to someone who has been abused? Someone who is depressed or suicidal?

Digital Church Network has seen a higher percentage of bi-vocational people operating in virtual reality churches than ever before. Metaversal cultural shifts will continue to lead us toward decentralization, which will continue to empower bi-vocational ministers to do ministry. But how are we equipping those people to handle such extreme situations? Arguably, many of our *physical* church pastors do not even know how to handle these extreme situations (as seminary hasn't prepared them). What does it look like to equip pastors, planters, and everyday people to handle trauma in virtual reality?

In Defense of Secondhand Trauma

While it's of extreme importance for us to meet the needs of people in virtual reality (even those under extreme duress or trauma), we cannot sacrifice ourselves in this process. Through Digital Church Network we have seen several examples of secondhand trauma. This may be a new concept to some.

People who hang around tobacco smokers experience what's called "secondhand smoke," as smoke from the first person (the smoker) is inhaled by a second person and ultimately impacts the health of that person negatively as well. Secondhand smoke can affect a non-smoker as tobacco smoke from the smoker invades the non-smoker's lungs.

Secondhand trauma can affect pastors, planters, and everyday people in much the same way. Depression can exude from one person and affect another. Meeting people where they are and pulling them out of dark places can put our virtual ministers in dark places that can have negative effects on their psyches and souls as well.

Secondhand trauma is not a reason to run from virtual reality ministry. Firefighters run *into* smoking buildings, not away from them. Yet firefighters wear suits that protect them from being burned, and they have equipment that allows them to rescue people safely. Of course, there will always be risks for firefighters, but they have trained and prepared as much as they possibly can. As ministers in virtual reality, how can we prepare ourselves (and others) for the attacks—even the secondary ones—that will come?

The Opportunity: Gear Up and Run into the Fire

It's easy to look at the complexities here, the mental health challenges we're facing in this ministry, and to run from the situation. Maybe virtual reality is unsalvageable. Maybe the skeptics are right, and the virtual realm is really beyond grace. However, God *did* lead Jonah into Nineveh. Paul *did* go to Rome. And I believe that God is calling many of us to reach a different type of person in the metaverse. The opportunity is not *away* from the fire but in it. We must go where God is calling us. Let's just be sure we have the right equipment and are prepared for the situation at hand.

(You've probably noticed by now that we've brought up more questions than answers. Through Digital Church Network, we will be tackling these issues in the months to come. Stay tuned to metaverse Church NEXT, and check out DCN's FAM Community to connect with other digital and metaverse ministers in this space.)

8

How to Get Started

AS YOU'RE READING THIS, I expect that one of two things is happening. By this point, either: 1) you've come to the personal realization that I'm a crackpot, and you're only finishing this book because of a bet, or 2) you're intrigued by the idea of church in virtual reality, and you're wondering what's next.

If you're in the "he's a crackpot" group, congratulations! The book is almost over and you're that much closer to claiming your prize. But if you're in the "I'm intrigued" group, let's go! As we get into the details, let's start by comparing the steps to planting a church in virtual reality to planting a physical church.

If your church is interested in investing in a new community or neighborhood by planting a new church, or even a new multi-site location, people are likely going to start asking questions about the community you are going into. Probably someone will talk about demographics to find neighborhoods that, on paper, align with your church. But remember, in virtual reality we have an opportunity to reach a different type of person. Don't automatically think "sameness."

DJ Soto, Lead Pastor of VR MMO Church, recommends a three-step process to help churches discover and solidify their ministry in the metaverse:

Step 1: Explore

Get that headset on and explore virtual reality worlds and communities. Meet people. Dialogue. Have shared experiences with them. Look (and pray for) your person of peace. Quite literally prayer-walk through virtual reality. See where God leads you as you're exploring strange new worlds.

Find people in your church who are excited about virtual reality. Be transparent with them. Ask questions and listen. God may have already equipped people in your church for this specific ministry. Maybe your job is to be patient and listen for God's leading, releasing control to these people. Maybe God will break you of your idea of what ministry is and will give you a new wineskin as you discover the new wine of virtual reality church. You're not going to know until you put on a headset.

Now that you have that headset, what do you do? As much fun as playing *Beat Saber* or *Population: One* is in virtual reality, I would challenge you to not just explore shared experiences but to walk around the communities and talk with people. Sit around someone's fireplace in AltspaceVR and have a conversation. Talk with the weirdest-looking avatar you can find in VRchat. Party it up in a dorm room in RecRoom. Eventually you will stop seeing the pixels of the avatars, and you'll experience the phenomenon that often happens in virtual reality ministry—you'll start to see these avatars as actual people.

If your exploration is rewarded by a calling to reach and disciple people in the metaverse, then let's move more aggressively to Step 2.

Step 2: Experiment

What will your connection look like in virtual reality? Is it more structured, like the physical church service, or is it more organic, like church in a VR bar? What will the culture of your church support? What is the role of volunteers? What does discipleship look like?

Start small and experiment. Get input from others. Release control where possible. Be OK with failure, which gets you one step closer to succeeding—not one step closer to quitting. Make the commitment to reiterate, to try again.

Remember that culture is shifting because of the metaverse. It will be very difficult to develop a five-year forecast that predicts the destination and route of the Church in the metaverse. Experimenting gives your church a better understanding in the short-term to make long-term decisions. Experimenting also gives your church an opportunity to take a risk and to empower different voices to shape your strategies.

Many people view experimenting as a weakness—and maybe it *is* a weakness. But assuming the posture of humility and learning how the metaverse operates will position us in a place where God can use us in ways that bring glory to him. James 4:6 says, "But he gives us more grace. That is why Scripture says: 'God opposes the proud but shows favor to the humble.'"

Step 3: Establish

Hopefully you are learning from the successes (and failures) of experiments, and you're moving closer to understanding the long-term potential for ministry. Some churches will work very hard to align virtual reality ministry with ministry that happens in physical space. There is beauty in seeing things centralized, in seeing commonality among ministries.

To be the bearer of bad news, though, sometimes God calls us down roads that don't make sense, and the opportunities of virtual reality may not always align under a singular strategy. Remember, if we're trying to reach different people, then our church may need to literally do something different. This is where the beauty of decentralization kicks in. Our virtual churches should have freedom to explore something different without being limited to doing the same thing as the physical church.

If we experimented properly, we're hopefully seeing a reproducible model of "church" in virtual reality. So, is the virtual community a church? Or is it a discipleship community? Or a missional opportunity?

The ecclesiological conversations centered around the metaverse are hugely important. They're also very slow. In many ways, we can't really talk ecclesiology until the technology solidifies, which is going to take a minute or two. (Truthfully, augmented reality and mixed reality will probably need to solidify before the Church can truly address the ecclesiological concern.)

Does this mean that the Church should ignore the virtual reality expression of church for now? No way! These experiments are necessary! But realize that even words like *establish* are still couched in an experimental mode.

While there may be some discrepancy in whether this virtual reality opportunity is truly a church, there can be no question as to the missional implications in this space. The mission field is ripe for the harvest. Lean into the opportunities in front of you, and let's worry later about what we're calling this.

(Digital Church Network is working with pastors, planters, and everyday people in starting metaversal churches and discipleship movements. If you're looking for ways to establish churches and movements in the metaverse, swing over to http://fam.digitalchurch.network and get up-to-date information on establishing churches in virtual reality and other metaverse spaces.)

9

Remember, Virtual Reality Is Only the Beginning

AT THIS POINT, we've talked about the different ideas surrounding church in virtual reality. We've talked theology and ecclesiology. We've discussed the opportunities and the horror stories. We've shown examples of successful churches in virtual reality, and we've heard from people who are regularly doing ministry in these virtual worlds. And, as much fun as churches in virtual reality are, we are merely scratching the surface of what the Church in the metaverse will look like. Augmented and mixed reality will blur the lines between physical reality and virtual reality, combining them into one (most likely complex) reality. Blockchain and cryptocurrencies will continue to disrupt organizational structures and shift toward decentralization, empowering individuals with the tools and accountability of major organizations. The metaverse is far more than a VR headset and avatars.

In 2023, Leadership Network will continue to tell the story of the Church in the metaverse, drawing attention to the ways that the Church needs to influence the metaverse and describing how the metaverse, in turn, needs to influence the Church. Leadership Network will continue to focus on Web3 technologies, and through content and communities, it will work to help churches embrace the metaverse. If your church is looking to take next steps to not just understand the metaverse but to experiment with the metaverse, consider joining one of Leadership

Network's Learning Communities, where we will continue to explain virtual reality as well as augmented reality, blockchain, cryptocurrencies, DAOs, NFTs, and more. For more information on our learning communities, check out http://leadnet.org/meta-lc.

Diversification, Decentralization, and Contextualization, Oh, My!

Church, COVID-19 has shifted, and the metaverse will continue to shift the physical (and virtual) worlds around us. To reach different people, our strategies need to shift.

If there's one takeaway from this book, it's the importance of embracing an experimental culture. Our churches have moved beyond simple ideology. Transitions toward diversification, decentralization, and contextualization are very possible in churches like yours, but they require an experimental attitude, which comes from a humble heart. For our churches to reach different people, we must experiment with different ideas. So, grab that headset, start prayer-walking through virtual reality, and see where God leads you.

Whether your church is exploring virtual reality, prayer-walking to find that person of peace, experimenting with a microchurch model infused with blockchain/DAO technology, or you're really excited to see how augmented reality is going to connect you to your digital church globally, we will only be able to take advantage of these opportunities if we have that heart to experiment, to try something different . . . to reach that different person.

ABOUT THE AUTHOR

In June 2000, Jeff Reed led his first online Bible study, taking 75 people around the planet through the book of James using a text-based system called Ultimate BB. He was doing digital ministry way before it was cool.

Since founding THECHURCH.DIGITAL in 2018, Jeff's passions have evolved into helping churches (and individuals too!) find their calling through digital discipleship, releasing people on digital mission, and planting multiplying digital churches. This pursuit is realized through DigitalChurch.Network, an organic, decentralized network for digital expressions of church, globally.

Jeff also champions metachurch development as the director of Metaverse Church Next for Leadership Network and works closely with NewThing Network, Exponential, and other globally facing, multiplication-friendly, gospel-centric organizations.

Jeff married his high-school sweetheart, Amy, and has two kids and two dogs. They live in Miami, Florida.

ENDNOTES

1 "History Of Virtual Reality," Virtual Reality Society, https://www.vrs.org.uk/virtual-reality/history.html.

2 Christo Petrov, "45 Virtual Reality Statistics That Rock the Market in 2022," Techyury.net, updated August 20, 2022, https://techjury.net/blog/virtual-reality-statistics/.

3 "New Report: 19% of U.S. Adults Have Tried VR," AR Insider, May 20, 2022, https://arinsider.co/2020/05/20/new-report-19-of-u-s-adults-have-tried-vr/.

4 Petrov, "45 Virtual Reality Statistics."

5 "Create with VR Grant Awarded to School of Design," Louisiana Tech University, August 9, 2022, https://www.latech.edu/2022/08/09/create-with-vr-grant-awarded-to-school-of-design/.

6 Joe Gvora, "Google Glass: What Happened to the Futuristic Smart Glasses?," Screen Rant, updated June 15, 2022, https://screenrant.com/google-glass-smart-glasses-what-happened-explained/.

7 Hartley Charleton, "Apple Planning to Replace the iPhone With AR Headset in 10 Years," Mac Rumors, December 1, 2021, https://www.macrumors.com/2021/12/01/apple-planning-to-replace-the-iphone-in-10-years/.

8 Brian Dean, "iPhone Users and Sales Stats for 2022," Backlinko, updated May 28, 2021, https://backlinko.com/iphone-users#iphone-key-stats.

9 Jack Purcher, "An Apple Patent Reveals What SharePlay AR could bring to a Future Mixed Reality Headset via a Plurality of Diorama-Views," Patently Apple, October 7, 2021, https://www.patentlyapple.com/2021/10/an-apple-patent-reveals-what-shareplay-ar-could-bring-to-a-future-mixed-reality-headset-via-a-plurality-of-diorama-views.html.

10 "What Is Blockchain Technology?," IBM, https://www.ibm.com/topics/what-is-blockchain.

11 "International Business Machines" (IBM) Company Profile, Fortune, https://fortune.com/company/ibm/.

12 "Meta," Wikipedia, https://en.wikipedia.org/wiki/Meta.

13 PJ Grisar, "Does Mark Zuckerberg know what Meta means in Hebrew?", Forward, October 29, 2021, https://forward.com/culture/477390/meta-dead-hebrew-facebook-mark-zuckerberg-new-name-metaverse/.

14 TeeVees Greatest, "Star Trek: The Original Series 1966–1969 Opening and Closing Theme," YouTube, April 27, 2016, https://www.youtube.com/watch?v=LIQsrvW6Ji4.

15 Jeff Reed, "Podcast 014: DJ Soto & Planting a Church in Virtual Reality," *THECHURCH.DIGITAL Podcast*, June 24, 2019, https://be.thechurch.digital/blog/podcast-014-dj-soto-planting-a-church-in-virtual-reality.

16 Jeff Reed, "Podcast 103: Jason Poling & The Virtual Reality Multisite Campus," *THECHURCH.DIGITAL Podcast*, November 9, 2020, https://be.thechurch.digital/blog/podcast-103-jason-poling-the-virtual-reality-multisite-campus.

17 *We Met in Virtual Reality*, directed and written by Joe Hunting (2022), released on HBO and HBOMax on July 27, 2022.

18 Hyun-Woo Lee, Sang Hun Kim, and Jun-Phil Uhm, "Social Virtual Reality (VR) Involvement Affects Depression When Social Connectedness and Self-Esteem Are Low: A Moderated Mediation on Well-Being," *Frontiers in Psychology*, November 30, 2021, https://www.frontiersin.org/articles/10.3389/fpsyg.2021.753019/full.

19 Helen Keller, BrainyQuote.com, https://www.brainyquote.com/quotes/helen_keller_131864.

20 David Malki, "True Stuff: Monk vs. the Printing Press," Wondermark®, January 13, 2011, http://wondermark.com/true-stuff-monk-vs-press/.

21 *In Praise of Scribes*, Amazon.com, https://www.amazon.com/Praise-Scribes-Laude-Scriptorum/dp/0872910660.

22 Chris Anderson, "Gutenberg and Luther," Church Works Media, October 6, 2017, https://www.churchworksmedia.com/2017/10/06/gutenberg-and-luther/.

23 "Luther and the Printing Press," Condordia University website, previously accessed at https://www.concordia.edu/blog/archive2017/Luther-and-the-printing-press-mcintosh-doty.html [broken].

24 Victor Hugo, *The Hunchback of Notre Dame*.

25 Jeff Reed, "Podcast 103."

26 Jeff Reed, "Podcast 104."

27 Calvary VR website: https://calvaryvr.com/.

28 Jay Kranda, "03: Three Digital Ministry Approaches," JayKranda.com, March 15, 2021, https://www.jaykranda.com/blog/2021/3/15/03-three-digital-ministry-approaches.

29 "Edgar Dale," Wikipedia, https://en.wikipedia.org/wiki/Edgar_Dale.

30 Chris Kolmar, "23 Amazing Virtual Reality Statistics [2022]: The Future of VR + AR," Zippia, April 19, 2022, https://www.zippia.com/advice/virtual-reality-statistics/.

31 Adam Ormord, "It's Time for a Relational Reformation," Grafted Life Ministries, https://www.graftedlife.org/articles/its-time-for-a-relational-reformation.

32 Jason Poling, "Theology of the Metaverse," May 4, 2022, Cornerstone Church Yuba City, http://cornerstoneyc.com/metaversetheology.

33 Poling, "Theology of the Metaverse."

34 Internet search of the phrase "pioneers get shot settlers get rich," Google.com.

35 Taylor Swift, with Max Martin and Shellback, "Shake It Off," released on August 19, 2014, by Big Machine Records.

36 Jeff Reed, "Podcast 130: Pastor SouZy - Understanding the Twitch, Streaming, Video Game Culture," *THECHURCH.DIGITAL Podcast*, May 16, 2021, https://be.thechurch.digital/blog/podcast-130-pastor-souzy-understanding-the-twitch-streaming-video-game-culture.

37 Jeff Reed, "Episode 193: Jason Poling & The Multiplying META Church," *THECHURCH.DIGITAL Podcast*, December 6, 2021, https://be.thechurch.digital/blog/episode-193-jason-poling-the-multiplying-meta-church.

38 Jay Kranda, "Sliding Scale of Online Offline Engagement" [Online Church], JayKranda.com, November 17, 2019, https://www.jaykranda.com/blog/2019/11/17/sliding-scale-of-online-offline-engagement.

39 Jeff Reed, "Episode 193."

40 Jeff Reed, "Podcast 130."

41 Mahatma Gandhi, Goodreads.com, https://www.goodreads.com/quotes/22155-i-like-your-christ-i-do-not-like-your-christians.

42 "RT 17: Christ-Centered Recovery with Nate Graybill," THECHURCH.DIGITAL, YouTube, August 17, 2022, https://www.youtube.com/watch?v=0GFvDq4ntYI.

HAVE A *CRAZY IDEA** FOR THE KINGDOM? LET'S TURN THAT IDEA INTO *REALITY*!

Made in the USA
Columbia, SC
17 November 2022